A SHORT
INTRODUCTION
TO MODAL LOGIC

CSLI
Lecture Notes
No. 30

A SHORT INTRODUCTION TO MODAL LOGIC

Grigori Mints

CENTER FOR THE STUDY
OF LANGUAGE
AND INFORMATION

Contents

Introduction

Modal logic is sometimes characterized as the logic of necessity and possibility, of "must be" and "may be." The aim of this book is to present both semantical and syntactical features of the subject and to illustrate them by detailed analysis of the three most popular modal systems S5, S4 and T. The text presupposes some knowledge of classical logic, although the necessary information is briefly recapitulated. Otherwise the material is self-contained. We concentrate on the logic side of the subject and provide philosophical motivations to show the point of the formal work itself. Most attention is paid to developments which took place after the first third of the 20th century. Historical topics are only touched upon.

This book partially follows the composition of Part I of Hughes and Cresswell 1968, but simplifies some technical aspects by stressing the use of semantic notions and related Gentzen-type methods. So our first modal system is S5, which is seen as a reformulation of the monadic predicate logic. An axiomatic treatment, which is sometimes a source of difficulty for beginners, is postponed to the last chapter. The systematic use of the Gentzen-type approach via refutation procedures also makes this course a suitable introduction to the proof theory of modal logic.

Webster's Dictionary[1] explains the meaning of some of our notions in the following way.

Mode. (Logic) The form in which a proposition connects the predicate and subject, whether by simple, contingent or necessary assertion; the form of the syllogism, as determined by the quantity and quality of the constituent propositions; mood.

Modal. (Logic) Indicating, or pertaining to, some mode or modality; expressing modality.

Modality. (Logic) That qualification of propositions according to which they are distinguished as asserting (or denying) the possibility, impossibility, contingency, or necessity of their content. A modal relation or quality; a mode or point of view under which an object presents itself to the mind. According to Kant, the quality of propositions, as assertory, problematical, or apodictic.

We shall see how these notions are made precise in the framework of the philosophical logic.

The words *necessary* and *possible* as used in modal logic include reference not only to actual state of affairs, but also to possible states or worlds. A statement A is true if it holds in the actual or present state of affairs; A is *necessarily true* (written $\Box A$ and read "necessary A") if it is true in every possible state of affairs. In this respect the *necessity* operator $\Box$ is similar to the universal quantifier $\forall s$ (for every s) in the familiar predicate logic. Another modal operation of *possibility* is related to existential quantifier $\exists$. Statement A is possible if it is true in some possible state of affairs. This situation is described by writing $\Diamond A$ which is read "possible A." By adding standard logical connectives such as $\neg$ (not), $\lor$ (or), $\rightarrow$ (implies), $\land$ (and), $\leftrightarrow$ (iff), one can describe more complicated modal notions. "*Impossible A*" is expressed by $\Box \neg A$ and means that A does not hold in any possible state of affairs. "*Contingent A*" is expressed by $\neg \Box A \land \neg \Box \neg A$ and means that neither A nor $\neg A$ is necessary. That is, A is neither necessary nor impossible, requiring that

---

[1] *The New International Dictionary of the English Language*, Second Edition, Merriam Company, Springfield MA, 1952.

there is a state where A is true and another state where A is false.

Another important modal notion is that of *strong implication* (which is sometimes called entailment). To say that a proposition p strongly implies q means that it is necessary that p implies q. We will use the symbol $\prec$ to denote strong implication. Then, $q \prec p$ is the same as $\Box(p \to q)$.

The systems of modal logic considered in this course are based on the familiar (*classical*) *propositional logic*. This latter system completely describes the logic of Boolean connectives, i.e., ones with operate with two *truth values*, true and false, often denoted by 1 and 0 respectively. It is because of this completeness that there is only one Boolean propositional logic. In the modal case, various systems of logic are possible. The original motivation for this diversity, particularly in the works of C. I. Lewis, was mainly syntactical. The aim there was to analyze interrelations between modal operators. Semantic aspects of modal logic became important later. The original motivation here was an analysis of the possibility operator in terms of the concept of *possible worlds* which is used to explain necessity and possibility of sentences.

These syntactic and semantic analyses of modal notions mentioned above provide a pattern for the analysis of other notions dealing with possible states of affairs, such as knowledge, belief, and temporal notions like 'always'. There are also close connections with topics where the notion of possible worlds enters in different ways, such as in intuitionistic logic, provability logic, and even modularity in logic programming. All of these systems are referred to as non-classical logics, to distinguish them from the familiar classical logic which deals with truth in one actual world or state of affairs.

Acknowledgments

The author wishes to thank the Department of Philosophy at Stanford University for the opportunity to present this material in a course in the Fall quarter of 1990. Also thanks to Dikran Karagueuzian for suggesting the publication of this book, and his help during the whole process of its preparation; to Tom Burke and Kaija Lewis for editing the manuscript and improving the language; and to Pilvi Veeber for typing an intermediate version.

1

Classical Propositional Logic

All modal systems we deal with are based on the *classical propositional logic*, or *classical propositional calculus*, to be abbreviated PC.

Let us recall some facts about PC.

1.1 Syntax

Well-formed formulas (or formulas, for short) of this system are constructed in the standard way from *propositional variables* or *propositional letters* denoted by

$$p,\ q,\ r,\ldots,p_1,\ q_1,\ r_1,\ldots$$

by means of logical connectives $\neg$ (negation, "not") and $\vee$ (disjunction, "or"). So p, $\neg q$, $(p \vee \neg p)$, $p \vee (q \vee r)$ etc., are formulas (while $p\neg\vee$ isn't).

1.2 Semantics: Truth Tables

Admissible values for propositional variables in the standard semantics for PC are *true* and *false*, often denoted by $1, 0$. *Truth-values* of compound formulas are computed from truth-values of variables by the standard rules summarized by the following truth-tables.

p	$\neg p$
1	0
0	1

p	q	$p \vee q$
1	1	1
1	0	1
0	1	1
0	0	0

The table for $\vee$ can be given more succinctly as follows:

$\vee$	1	0
1	1	1
0	1	0

So every given assignment of truth values to variables occuring in a given formula determines the truth-value of this formula. Consider some examples.

Example 1.1 Let $\alpha \equiv \neg p \vee p$, i.e., α stands for $\neg p \vee p$. Then for $v(p) = 1$ we have $v(\neg p) = 0$ and $v(\neg p \vee p) = 1$. The truth values of p, $\neg p$, and α, respectively, can be tabulated as follows.

p	$\neg p$	$\neg p \vee p$
1	0	1
0	1	1

So formula α is true, i.e., takes value *true* under every assignment of truth values. This means by definition that it is a *tautology* or a *valid formula* of PC.

Example 1.2 $\alpha \equiv \neg(p \vee q) \vee p$. Let $v(\beta)$ denote the truth value of a formula β under a given assignment v. Then, consider the assignment where $v(p) = 0$ and $v(q) = 1$. Then $v(\alpha) = v(\neg(p \vee q)) \vee 0 = v(\neg(p \vee q)) = v(\neg(0 \vee 1)) = v(\neg 1) = 0$.

Thus F is false under a given assignment, so it is not a tautology. The assignment $v(p) = 0$, $v(q) = 1$ is said to be a *falsifying assignment* for α. Assignment $v(p) = v(q) = 1$ gives $v(\alpha) = 1$ and so is a *verifying* (or *satisfying*) *assignment*.

Since operators $\neg$ and $\lor$ act on truth-values of their arguments, i.e., are truth-functions, they are called *truth-functional operators*.

An operator with one argument (like $\neg$) is called *monadic,*; an operator with two arguments (like $\lor$) is said to be *dyadic*.

Further truth-functional operators can be defined in terms of $\neg$, $\lor$. In our setting this will mean that we introduce them as abbreviations. The *definitions* are:

$$\alpha \land \beta \quad \equiv \quad \neg(\neg\alpha \lor \neg\beta) \tag{1.1}$$
$$(\alpha \to \beta) \quad \equiv \quad \neg\alpha \lor \beta$$
$$(\alpha \leftrightarrow \beta) \quad \equiv \quad ((\alpha \to \beta) \land (\beta \to \alpha))$$

where α, β represent any formulas of PC. Here $\land$ is called the conjunction sign (read "and"), $\to$ is the implication sign (read "implies" or "if ... then ...") and $\leftrightarrow$ is the equivalence sign (read "equivalent to" or "if and only if"). To understand the reason for that, let us compute the truth-tables according to definition (1.1). The results are:

$\land$	1	0		$\to$	1	0		$\leftrightarrow$	1	0
1	1	0		1	1	0		1	1	0
0	0	0		0	1	1		0	0	1

In other words,

- $\alpha \land \beta$ is true iff both α *and* β are true;
- $\alpha \leftrightarrow \beta$ is true iff the value of α is equal to the value of β, i.e., iff α and β are equivalent; and
- $\alpha \to \beta$ is true iff the truth of α implies the truth of β.

In the definitions and proofs we sometimes will indicate modifications which allow us to treat connectives different from $\lor, \neg$ as primitive, which often shortens computations.

We can also state the truth-table definitions of these connectives in abbreviated linear form. First of all we have

$$v(\neg\alpha) \;=\; 1 - v(\alpha) \tag{1.2}$$

which is easily checked against the truth-table: $v(\neg 0) = 1 = 1 - 0$ verifies (1.2) for $\alpha = 0$, and $v(\neg 1) = 0 = 1 - 1$ does this for $\alpha = 1$.

The table for $\vee$ has only one zero value for the case when both arguments are zero. This gives:

$$
\begin{aligned}
v(\alpha \vee \beta) &= \max(v(\alpha), v(\beta)) \\
v(\alpha \vee 0) &= v(\alpha) \\
v(\alpha \vee 1) &= 1
\end{aligned}
\tag{1.3}
$$

Similarly from the table for $\wedge$ we have:

$$
\begin{aligned}
v(\alpha \wedge \beta) &= \min(v(\alpha), v(\beta)) \\
v(\alpha \wedge 0) &= 0 \\
v(\alpha \wedge 1) &= v(\alpha)
\end{aligned}
\tag{1.4}
$$

The table for $\leftrightarrow$ gives:

$$
\begin{aligned}
v(\alpha \leftrightarrow \beta) &= 1 \ \text{ iff } \ v(\alpha) = v(\beta) \\
v(\alpha \leftrightarrow 0) &= v(\neg\alpha) \\
v(\alpha \leftrightarrow 1) &= v(\alpha)
\end{aligned}
\tag{1.5}
$$

Finally, $v(\alpha \to \beta) = 0$ iff $v(\alpha) = 1$, $v(\beta) = 0$. So in particular,

$$
\begin{aligned}
v(\alpha \to \beta) &= \max(1 - v(\alpha), v(\beta)) \\
v(1 \to \beta) &= v(\beta) \\
v(0 \to \beta) &= 1
\end{aligned}
\tag{1.6}
$$

The latter of the relations in (1.6) is read: "contradiction (or false) implies everything." This is an important feature of the truth-functional operator $\to$. It is called *material implication* to stress this feature.

Let us formulate again the truth conditions for the connectives $\vee$, $\wedge$, $\rightarrow$, and their "negations."

$$v(\alpha \vee \beta) = 1 \quad \text{iff} \quad v(\alpha) = 1 \text{ or } v(\beta) = 1 \qquad (1.7)$$

$$v(\neg(\alpha \vee \beta)) = 1 \quad \text{iff} \quad v(\neg\alpha) \text{ and } v(\neg\beta) = 1 \qquad (1.8)$$

$$v(\neg\neg\alpha) = v(\alpha) \quad \text{since} \quad v(\neg\neg\alpha) = 1 - v(\neg\alpha) \qquad (1.9)$$

$$= 1 - (1 - v(\alpha)) = v(\alpha)$$

$$v(\alpha \rightarrow \beta) = 1 \quad \text{iff} \quad v(\neg\alpha) = 1 \text{ or } v(\beta) = 1 \qquad (1.10)$$

$$v(\neg(\alpha \rightarrow \beta)) = 1 \quad \text{iff} \quad v(\alpha) = 1 \text{ and } v(\neg\beta) = 1 \qquad (1.11)$$

$$v(\alpha \wedge \beta) = 1 \quad \text{iff} \quad v(\alpha) = v(\beta) = 1 \qquad (1.12)$$

$$v(\neg(\alpha \wedge \beta)) = 1 \quad \text{iff} \quad v(\neg\alpha) = 1 \text{ or } v(\beta) = 1 \qquad (1.13)$$

Relations (1.7)–(1.13) allow us to test formulas for the existence of satisfying assignments of truth values. Consider some examples.

Example 1.3 $\alpha \equiv (p \wedge q) \wedge \neg r$. Suppose v is a satisfying assignment for α, i.e., $v(\alpha) = 1$. Then by (1.12) $v(p \wedge q) = 1$ and $v(\neg r) = 1$, and again by (1.12) $v(p) = v(q) = 1$. Now we have found a satisfying assignment $v(p) = v(q) = 1$, $v(r) = 0$, the latter being an equivalent of $v(\neg r) = 1$.

We can write this in the form of a tree, by putting the original formula α into the bottom node of the tree and letting the tree grow upwards. In this case the tree consists of exactly one branch which is obtained working bottom-up from the lowermost node.

satisfying assignment: $v(p) = v(q) = 1; v(r) = 0$

$$\begin{array}{ccc} 1 & 1 & 0 \end{array}$$

$p, \quad q, \quad \neg r \quad$ by (1.12)

$p \wedge q, \quad \neg r \quad$ by (1.12)

$(p \wedge q) \wedge \neg r \quad$ original formula

How do we verify that the assignment in fact satisfies the original formula? One can substitute values for variables and compute. A more fruitful approach is to note that relation (1.12) is an equivalence relation, i.e., it works in both directions. So every assignment which verifies all formulas at the upper line must verify all formulas downward.

Example 1.4 $\alpha \equiv p \wedge \neg(r \vee q)$. We have the tree:

satisfying assignment: $v(p) = 1; v(q) = v(r) = 0$

1 0   0

$p, \; \neg r, \;\;\; \neg q$ by (1.8)

$p, \; \neg(r \vee q)$ by (1.12)

$p \wedge \neg(r \vee q)$

Example 1.5 $\alpha \equiv p \wedge (\neg p \vee q)$. Here the tree is branching, due to the "or" condition in (1.7). The left branch is contradictory and does not produce a verifying assignment.

verifying assignment: $v(p) = v(q) = 1$

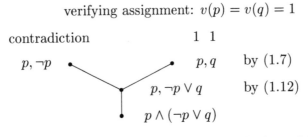

contradiction   1 1

$p, \neg p$ p, q by (1.7)

$p, \neg p \vee q$ by (1.12)

$p \wedge (\neg p \vee q)$

Example 1.6 $\alpha \equiv (p \wedge q) \wedge \neg(q \wedge p)$. The tree looks as follows:

contradiction   contradiction

$p, q, \neg q$ $p, q, \neg p$ by (1.13)

$p, q, \neg(q \wedge p)$ by (1.12)

$p \wedge q, \neg(q \wedge p)$ by (1.12)

$(p \wedge q) \wedge \neg(q \wedge p)$

Here all leaves (uppermost nodes of the tree) contain contradictions, and there is no satisfying assignment for the formula

α considered. This means

$$v(\alpha) \;=\; 0 \tag{1.14}$$

for any valuation v, and so for all v

$$v(\neg\alpha) \;=\; 1. \tag{1.15}$$

1.3   The Refutation Procedure

This latter trick allows us to use the tree procedure to test formula for being tautologies.

In short, a formula α is a tautology iff $\neg\alpha$ is identically false, i.e., $v(\alpha) = 0$ for all assignments v. Indeed $v(\neg\alpha) = 1 - v(\alpha)$, so $v(\alpha) = 1$ means that $v(\neg\alpha) = 0$.

Procedure $\mathcal{P}$. To test whether formula α is a tautology, form $\neg\alpha$ and test it for a satisfying assignment. We say α is a tautology when the former test gives a negative answer, i.e., if all branches of the tree end in contradictory leaves.

An exact formulation of this procedure and the proof of its correctness (of the pronouncement it makes) will be given after the following example.

Example 1.7 $\alpha \equiv (p \vee (q \wedge r)) \to (p \vee q)$. α is a tautology, as proved by the following tree.

contradiction

contradiction	$q, r, \neg p, \neg q$	by (1.12)
$p, \neg p, \neg q$	$(q \wedge r), \neg p, \neg q$	by (1.7)
	$p \vee (q \wedge r), \neg p, \neg q$	by (1.8)
	$p \vee (q \wedge r), \neg(p \vee q)$	by (1.11)
	$\neg((p \vee (q \wedge r)) \to (p \vee q))$	

A description of the procedure $\mathcal{P}$ for finding a satisfying assignment by construction of a refutation tree, will be given for finite sets of formulas. Assignments are extended from formulas

to lists of formulas by putting

$$v(\Gamma) = v(\alpha_1) \wedge \ldots \wedge v(\alpha_n)$$

for $\Gamma \equiv \alpha_1, \ldots, \alpha_n$. So

$$v(\Gamma) = 1 \quad \text{iff} \quad v(\alpha_1) = 1 \wedge \ldots \wedge v(\alpha_n) = 1.$$

The procedure is specified by stages as follows.

Stage 0: Place the given set of formulas Γ at the bottom of the tree to be constructed. Go to stage 1.

Stage $N + 1$. Consider the leaves (uppermost nodes) of the tree constructed at stage N. Different steps are now possible, depending on what occurs at these nodes.

Step $\mathcal{P}^0$. If each of these leaves contains an explicit contradiction, i.e., a pair of formulas p, $\neg p$ for some variable p, then the procedure is *finished* with a negative answer: the original set Γ is *contradictory*, i.e., it is not satisfied by any assignment.

If there is a leaf which does not contain an explicit contradiction, take one of these, say Γ, and perform the following actions as appropriate.

Step $\mathcal{P}^{\neg\neg}$. If Γ contains a formula of the form $\neg\neg\alpha$, then extend the current leaf to a new node where this formula $\neg\neg\alpha$ is replaced by α, and go to the next stage.

Φ, α

$\Phi, \neg\neg\alpha$

Step $\mathcal{P}^{\neg\vee}$. If Γ contains a formula of the form $\neg(\alpha \vee \beta)$, then replace it at a new leaf by the pair $\neg\alpha$, $\neg\beta$, and go to the next stage.

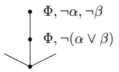

Step $\mathcal{P}^\vee$. If Γ contains a disjunction $(\alpha \vee \beta)$, then extend the current leaf of the tree by two new leaves; one with $\alpha \vee \beta$ replaced by α, and the second with $(\alpha \vee \beta)$ replaced by β; and then go to the next stage.

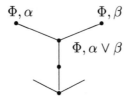

Step $\mathcal{P}^{val}$. If Γ consists only of variables and negations of variables, and is not contradictory, i.e., does not contain a pair $p, \neg p$, then exit from the procedure with the following verifying valuation:

$$v(q) = 1 \quad \text{iff} \quad q \in \Gamma.$$

This concludes the description of the search procedure $\mathcal{P}$ for our language with connectives $\neg, \vee$.

If $\wedge, \rightarrow$ are treated as primitive connectives one has to include the following steps for $\wedge, \rightarrow, \neg\wedge, \neg \rightarrow$:

$$\mathcal{P} \wedge \frac{\Phi, \alpha, \beta}{\Phi, \alpha \wedge \beta} \qquad \mathcal{P} \rightarrow \frac{\Phi, \neg\alpha; \ \Phi\beta}{\Phi, \alpha \rightarrow \beta}$$

$$\mathcal{P}\neg \wedge \frac{\Phi, \neg\alpha; \ \Phi, \neg\beta}{\Phi, \neg(\alpha \wedge \beta)} \qquad \mathcal{P}\neg \rightarrow \frac{\Phi, \alpha, \neg\beta}{\Phi, \neg(\alpha \rightarrow \beta)}$$

Lemma 1.1 *Procedure $\mathcal{P}$ always terminates and the refutation tree is finite.*

Proof. Let b be the number of binary connectives in Γ, and n be the number of negations. We prove by induction on $2b + n$

(to be denoted by p) that the tree constructed by the procedure $\mathcal{P}$ for Γ has at most p levels. If Γ consists only of propositional variables and their negations then the tree contains 0 levels and the statement is obvious. Otherwise the tree is constructed by one of the steps $\mathcal{P}^{\neg\neg}, \mathcal{P}^{\vee}, \mathcal{P}^{\neg\vee}$. In the first cast, b is preserved and n is changed to $(n-2)$, so the new value of p is $b+n-2$ and induction hypothesis gives required estimate.

At the step $\mathcal{P}^{\vee}$ the value of n is not increased and the value of b is decreased at least by 1, so p is again decreased at least by 2.

At the step $\mathcal{P}^{\neg\vee}$, the value of b is decreased by 1, but the value of n is increased by 1, so we have $2(b-1) + n + 1 = 2b + n - 1 < p$ as the new value. By the induction hyothesis, the number of levels of the tree for the upper node of $\mathcal{P}^{\neg\vee}$ is at most $2b + n - 1$, so the number of levels above the lower node is bounded by $2b + n - 1 + 1 = p$ as required.

To find an estimate for the number of nodes in the refutatiuon tree, note that this tree is binary, i.e., the number of predecessors of any node is at most 2. Hence the number of nodes n_{l+1} at the level $l + 1$ is at most twice the number n_1. This implies (by induction on l)

$$n_l \leq 2^{l+1}$$

and so the total number of nodes is bounded by

$$1 + 2 + \ldots + 2^{L+1} = 2^{L+2}$$

where L is the number of levels.   □

1.4   Soundness and Completeness

Theorem 1.2 *Procedure $\mathcal{P}$ is (a) sound and (b) complete. That is,*

(a) if $\mathcal{P}$ ends with an assignment v for the list of formulas Γ, then $v(\Gamma) = 1$; and

(b) if $v(\Gamma) = 1$ for some v, then $\mathcal{P}$ ends with such an assignment v.

Proof. (a) Assume that $\mathcal{P}$ has finished by discovering a satisfying assignment v at some leaf l of the refutation tree. Consider the path from that leaf down to the bottom node b.

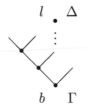

Let Γ be the original formula listed at the bottom node, and let Δ be the list at leaf l. We have

$$v(\Delta) = 1$$

by assumption and have to prove $v(\Gamma) = 1$. We prove the following lemma first.

Lemma 1.3 *Let c be a non-terminal node in the refutation tree, let c' (or c', c'') be all of its immediate predecessors, and let c, c', c'' contain lists of formulas Π, Π', Π'', respectively. Then, for any valuation v,*

$$v(\Pi) = 1 \quad \textit{iff} \quad v(\Pi') = 1 \textit{ or } v(\Pi'') = 1. \tag{1.16}$$

Note that when c has only one predecessor c' then the statement above takes the form

$$v(\Pi) = 1 \quad \textit{iff} \quad v(\Pi') = 1 \tag{1.17}$$

Proof. Let c be a non-terminal node of the tree, containing the list Π. Immediate predecessors of c are constructed at steps $\mathcal{P}^\vee$, $\mathcal{P}^{\neg\vee}$, $\mathcal{P}^{\neg\neg}$, etc. Each of these steps corresponds to exactly one of the relations (1.7), (1.8), (1.9), etc.

If the corresponding relation contains an "and" clause (like (1.8), (1.11), or (1.12)), then node c has only one predecessor c', where two formulas α, β are added to replace analyzed formula $\alpha \wedge \beta$, and we have

$$\Pi \equiv \Phi, \alpha \wedge \beta; \qquad \Pi' \equiv \Phi, \alpha, \beta :$$

$$
\begin{array}{ll}
c' \;\bullet & \Phi, \alpha, \beta \\
\; \Big| & \\
c \;\bullet & \Phi, \alpha \wedge \beta
\end{array}
$$

If $v(\Phi) = 0$ then $v(\Pi) = v(\Pi') = 0$ and (1.16) holds trivially. If $v(\Phi) = 1$, then $v(\Pi) = v(\alpha \wedge \beta); v(\Pi') = v(\alpha) \wedge v(\beta)$ and (1.16) follows from (1.8).

The relation (1.9) is treated even more easily. Here

$$\Pi \equiv \Phi, \neg\neg\alpha, \; \Pi' \equiv \Phi, \alpha$$

and (1.15) follows from $v(\neg\neg\alpha) = v(\alpha)$.

Now consider the case of extending the node c where the corresponding relation contains an "or" clause, namely (1.7). Then c has two immediate predecessors c', c'' and we have

$$\Pi \equiv \Phi, \alpha \wedge \beta; \quad \Pi' \equiv \Phi, \alpha; \quad \Pi'' \equiv \Phi, \beta$$

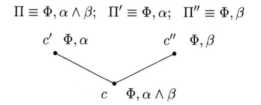

$$
\begin{array}{lll}
c' \;\; \Phi, \alpha & & c'' \;\; \Phi, \beta \\
& c \;\; \Phi, \alpha \wedge \beta &
\end{array}
$$

Again (1.16) is trivial for $v(\Phi) = 0$, and in the case of $v(\Phi) = 1$, it holds by (1.7). The lemma is proved. $\qquad\square$

If $\wedge, \to$, etc., are treated as primitive, then the steps corresponding to relations containing the "and" clause, (1.11) and (1.12), are treated like $\mathcal{P}\wedge$; and the "or" relations in (1.10) and (1.13) are treated like (1.7).

We now conclude the proof of clause (a) in the theorem. From (1.16) we have the following: if node c' is an immediate predecessor of c in the tree, and they contain lists Π, Π' respectively, then

$$v(\Pi') = 1 \quad \text{implies} \quad v(\Pi) = 1 \text{ for any } v. \qquad (1.18)$$

From this it follows that (1.18) holds also in the case when c' is an arbitrary (not necessarily immediate) predecessor of c, i.e., when they are in the same path in the tree. (Use mathematical induction on the number of nodes between c and c'.) Now take $c' = l$ and $c = b$, to conclude the proof of (a).

(b) First, note that the procedure terminates by the Lemma 1.1. Let v be an assignment such that $v(\Gamma) = 1$. Going bottom up through the tree, we distinguish at each level a node c such that $v(\Pi) = 1$ for the list Π placed at c. For the bottom level we have this by the assumption that $v(\Gamma) = 1$. To ascend from a node c already reached, apply (1.16) to choose the predecessor of c for which v is a satisfying assignment.

$$c'\ \Pi' \qquad c''\ \Pi'' \qquad v(\Pi') = 1 \text{ or } v(\Pi'') = 1$$

$$c\ \Pi \qquad\qquad\qquad v(\Pi) = 1$$

When the uppermost level of the tree is reached, one of the steps $\mathcal{P}^0$, $\mathcal{P}^{val}$ should apply, and it is exactly $\mathcal{P}^{val}$ which applies to the distinguished node since $v(\Pi) = 1$ for the list Π in this node. This completes the proof of Theorem 1.1. $\qquad\qquad\square$

By reformulating the steps of procedure $\mathcal{P}$ as inference rules, we obtain a complete and sound deduction system for PC. Consider the following axioms and rules.

Deductive System G.

○ Derivable objects are finite lists of formulas treated up to the order of formulas in them.

○ Axioms: $\Gamma, p, \neg p$ (for some letter p).

○ Inference rules:

$$\frac{\Gamma, \alpha}{\Gamma, \neg\neg\alpha}(\neg\neg) \qquad \frac{\Gamma, \neg\alpha, \neg\beta}{\Gamma, \neg(\alpha \vee \beta)}(\neg\vee) \qquad \frac{\Gamma, \alpha;\ \Gamma, \beta}{\Gamma, \alpha \vee \beta}(\vee)$$

$$\frac{\Gamma, \alpha, \beta}{\Gamma \alpha \wedge \beta}(\wedge) \qquad \frac{\Gamma, \neg\alpha \Gamma \neg\beta}{\Gamma, \neg(\alpha \wedge \beta)}(\neg\wedge)$$

$$\frac{\Gamma, \neg\alpha;\ \Gamma, \beta}{\Gamma, \alpha \to \beta}(\to) \qquad \frac{\Gamma, \alpha, \neg\beta}{\Gamma, \neg(\alpha \to \beta)}(\neg\to)$$

The rules other than the first three are present only when $\wedge$, $\to$ are added as primitive connectives.

Systems of this kind were first systematically studied by G. Gentzen.

The system G is designed to prove contradictions: if the premisses (subsequents above the line) are false, then the conclusion of the rule (subsequent below the line) is false, too.

Theorem 1.4 A list Γ of formulas is inconsistent (i.e., has no satisfying assignment) iff it is derivable in the system G.

Proof. By Theorem 1.1, the list Γ is inconsistent iff all leaves in the refutation tree for Γ are contradictory. But the latter property is so only when this refutation tree is a derivation of Γ in the system G. Indeed, all steps of the construction of the refutation tree are made in exact correspondence with the inference rules for G. The only remaining property of the derivation to be verified is that all the leaves are axioms, but this means (in the case of the system G) precisely that they contain a contradiction.   □

1.5   Substitution of Equivalents

We have already seen that the eqivalence connective $\leftrightarrow$ plays the role of equality of propositions. This impression is supported by the following observation:

$$(p \leftrightarrow q) \rightarrow (\alpha_r[p] \leftrightarrow \alpha_r[q]) \qquad (1.19)$$

is a tautology for any formula α, where $\alpha_r[\beta]$ is the result of substitution β for all occurences of r in α.

Indeed, take any valuation v. If $v(p) \neq v(q)$, then the formula in (1.19) is true under assignment v since its premise is false. If $v(p) = v(q)$, then $\alpha_r[p]$ and $\alpha_r[q]$ are evaluated in the same way by the assignment v, so $v(\alpha_r[p]) = v(\alpha_r[q])$ as required.

Using standard substitution and the rule of modus ponens, we first get from (1.20) the formula

$$(\beta \leftrightarrow \gamma) \rightarrow (\alpha[\beta] \leftrightarrow \alpha[\gamma]) \qquad (1.20)$$

for any formulas α, β, γ (where an indication of the variable r

to be substituted for is dropped), and then the inference rules

$$\frac{\beta \leftrightarrow \gamma}{\alpha[\beta] \leftrightarrow \alpha[\gamma]}(\text{Eq}) \qquad \frac{\beta \leftrightarrow \gamma;\ \alpha[\beta]}{\alpha[\gamma]}(\text{Eq})$$

which allow us to replace subformula β in a formula by any equivalent formula γ. In other words, if $\beta \leftrightarrow \gamma$ is a tautology then $\alpha[\beta] \leftrightarrow \alpha[\gamma]$ is a tautology. And if in addition $\alpha[\beta]$ is a tautology, then $\alpha[\gamma]$ is a tautology too. These rules will be referred to collectively as (Eq). Rules (Eq) allow us to make transitions like

$$\frac{\alpha[\beta \vee \gamma]}{\alpha[\gamma \vee \beta]} \qquad (\text{Comm}) \qquad \frac{\alpha[\beta \wedge \gamma]}{\alpha[\gamma \wedge \beta]}$$

$$\frac{\alpha[\beta \vee (\gamma \vee \delta)]}{\alpha[(\beta \vee \gamma) \vee \delta]} \qquad (\text{Assoc}) \qquad \frac{\alpha[\beta \wedge (\gamma \wedge \delta)]}{\alpha[(\beta \wedge \gamma) \wedge \delta]}$$

$$\frac{\alpha[\beta \vee (\gamma \wedge \delta)]}{\alpha[(\beta \vee \gamma) \wedge (\beta \vee \delta)]} \qquad (\text{Distr}) \qquad \frac{\alpha[\beta \wedge (\gamma \vee \delta)]}{\alpha[(\beta \wedge \gamma) \vee (\beta \wedge \delta)]}$$

and so forth.

1.6 The Refutation Procedure and Disjunctive Normal Form

We have seen that consistent paths in the refutation tree for a formula α (or, in our formulation rather, consistent leaves of that tree) correspond to truth assignments satisfying α. This connection will be clarified and extended later to more general situations, but now let us explain the connection with the well-known notion of disjunctive normal form.

For any list Γ of formulas, let $\bigwedge \Gamma$ denote the conjunction of all formulas in Γ, so that for $\Gamma \equiv \alpha_1, \ldots, \alpha_n$, we have $\bigwedge \Gamma \equiv (\alpha_1 \wedge \ldots \wedge \alpha_n)$. In particular, for $n = 0$ (corresponding to the empty list) we have $\bigwedge \Gamma = true$, i.e., some fixed formula of the form $p \vee \neg p$.

Theorem 1.5 *Let α be a formula, T its (not necessarily completed) refutation tree, and $l_1, \ldots, l_n$ the complete list of its* *leaves. If $\Gamma_1, \ldots, \Gamma_n$ are lists of formulas placed at $l_1, \ldots, l_n$ respectively, then*

$$\alpha \leftrightarrow (\wedge \Gamma_1 V \ldots V \wedge \Gamma_n) \tag{1.21}$$

is a tautology.

Note: One can take only consistent (non-contradictory) leaves.

To prove the theorem, we use the following lemma. For the sake of saving space, we will use the shorthand of a slash, as in

$$\Gamma'/\Gamma \text{ to indicate } \frac{\Gamma'}{\Gamma}.$$

Lemma 1.6 *All the rules of system* G *are equivalent transformations: if the rule has the form Γ'/Γ (one-premise rule), then $\wedge \Gamma' \leftrightarrow \wedge \Gamma$ is a tautology. If the rule has the form $\Gamma'; \Gamma''/\Gamma$ (two-premise rule), then*

$$(\wedge \Gamma' \vee \wedge \Gamma'') \leftrightarrow \wedge \Gamma \tag{1.22}$$

is a tautology.

Proof. Check all the rules using replacement of equivalents. For example, the treatment of the rule $(\neg\neg)$ uses equivalence $\neg\neg\alpha \leftrightarrow \alpha$, the rule $(\neg\vee)$ uses equivalence $\neg(\alpha \vee \beta) \leftrightarrow (\neg\alpha \wedge \neg\beta)$ and the rule $(\neg\wedge)$ uses equivalence $\neg(\alpha \wedge \beta) \leftrightarrow (\neg\alpha \vee \neg\beta)$. The latter rule has the form

$$\Gamma, \neg\alpha; \Gamma, \neg\beta/\Gamma, \neg(\alpha \wedge \beta)$$

and denoting $\wedge\Gamma$ by γ we have to prove

$$((\gamma \wedge \neg\alpha) \vee (\gamma \wedge \neg\beta)) \leftrightarrow (\gamma \wedge \neg(\alpha \wedge \beta)). \tag{1.23}$$

Replacing $\neg(\alpha \wedge \beta)$ by $\neg\alpha \vee \neg\beta$ we turn (1.23) into an instance of distributivity, which is a tautology. Lemma 1.4 is thus proved.

Proof of the theorem. We use induction on the stages of constructing a refutation tree, applying Lemma 1.6 up the tree. At stage 0 (induction base case) the tree consists of the bottom node where the formula α itself is placed, so (1.21) takes the form $\alpha \leftrightarrow \alpha$ and is obvious. To prove the induction step, assume that (1.21) is true at stage N and recall that stage $N + 1$

consists (unless it is final, in which case the tree is not changed) in extending one of the leaf nodes (say l_n, for definiteness) into one or two nodes (l'_n, or l'_n and l''_n). Consider just the second possibility. We are given (1.22) and have to prove the formula

$$\alpha \leftrightarrow (\wedge\Gamma_1 \vee \ldots \vee \wedge\Gamma_{n-1} \vee \wedge\Gamma'_n \vee \wedge\Gamma''_n).$$

This is obtained from (1.21) by substitution of equivalents (1.22).   □

Corollary 1.7 *Any formula α of* PC *is equivalent to*

$$V_i \wedge \Gamma_i \qquad\qquad (1.24)$$

when disjunction is taken over all (consistent) leaves Γ_i of its completed refutation tree. So (1.24) is a disjunctive normal form of α.

A more detailed analysis of the proof of this corollary shows that the steps of construction of the refutation tree are just standard steps of reduction of a propositional formula to disjunctive normal form.

2

Classical Monadic Predicate Logic

2.1 Syntax

Atomic formulas of this logic have the form $P(a)$ where P is a monadic (that is one-argument) *predicate letter* or *predicate variable* and a is an *individual* variable. *Well-formed formulas*, or simply formulas, are constructed from atomic ones by boolean connectives $\vee$, $\neg$ (and defined ones like $\wedge$, $\rightarrow$, $\leftrightarrow$) and quantifiers $\forall$, $\exists$. It is understood that there is an inexhaustible supply of predicate variables to be denoted by $P, Q, R, P_1, P_2, \ldots$ and of individual variables to be denoted by $a, b, c, a_1, a_2, \ldots, x, y, z, x_1, x_2, \ldots$

Example 2.1 The following are formulas:

$P(x)$,

$P(x) \rightarrow Q(x)$,

$\forall x (P(x) \rightarrow Q(x))$,

$\forall x (P(x) \rightarrow Q(x)) \rightarrow (\forall x P(x) \rightarrow \forall x Q(x))$,

$\exists x \forall y P(y) \rightarrow \forall y P(y)$

Recall that each of the quantifiers can be defined in terms of the remaining one,

$$\forall x A \leftrightarrow \neg \exists x \neg A;$$

$$\exists x A \leftrightarrow \neg \forall x \neg A$$

but we prefer to have both as primitives.

Occurences of an individual variable x in a formula can be *bound* (by quantifiers $\exists x$, $\forall x$) or *free* (if not in the *scope* of any such quantifier). In the process of substituting a variable y for another variable x in a formula $A(x)$, a *collision* can happen if some occurence of x in $A(x)$ is in the scope of a quantifier $\forall a$. In this case it is understood that the quantified variable a is *renamed*, i.e., the part $\forall a B(a)$ is replaced by $\forall z B(z)$ for a new variable z.

2.2   Semantics: Valuation Rules

A semantics for predicate logic is defined in terms of a *model*, which is of the form $M \equiv \langle D, V \rangle$, where D is non-empty set and V is a valuation of predicates in D. Namely, for any predicate letter P one has $V(P)\colon D \to \{0, 1\}$, i.e., for any $d \in \mathcal{D}$,

$$V(P)(d) \in 0, 1 \tag{2.1}$$

If $V(P)(d) = 1$, one says that $P(d)$ is *true in the model* M, and if $V(P)(d) = 0$ then $P(d)$ *is false in* M.

The *truth-value* $M(\alpha)$ of a formula α in a model M is defined only for constant formulas over M, that is, for objects of the form $\alpha(d_1, \ldots, d_n)$ where $\alpha(x_1, \ldots, x_n)$ is a formula with free variables $x_1, \ldots, x_n$ (exactly), and $A(d_1, \ldots, d_n)$ is the result of substituting objects $d_1, \ldots, d_n \in D$ for the variables $x_1, \ldots, x_n$. The definition of the truth-value of a formula (which we recapitulate for the comparison below) $M(\alpha)$ is given by induction on the number m of logic connectives $\neg$, $\vee$, $\forall$, $\exists$ in α.

The base case $m = 0$ is essentially given above:

$$M(P(d)) = V(P)(d) \tag{2.2}$$

The induction step for boolean connectives is obvious:

$$M(\neg \alpha) \;=\; 1 - M(A) \tag{2.3}$$
$$M(\alpha \vee \beta) \;=\; \max(M(\alpha), M(\beta)) = M(\alpha) \vee M(\beta) \tag{2.4}$$

where $\vee$ in $M(\alpha) \vee M(\beta)$ stands for the boolean disjunction (i.e., the maximum of truth values 0,1).

The quantifiers are treated as multiple $\wedge$ and $\vee$ extended over the individual domain D:

$$M(\forall x \alpha(x)) = \min_{d \in D} M(\alpha(d)) \qquad (2.5)$$

$$M(\exists x \alpha(x)) = \max_{d \in D} M(\alpha(d)) \qquad (2.6)$$

2.3 The Refutation Procedure

Let us write down relations for the truth of quantified formulas and their negations, which are similar to relations (1.7)–(1.13) of the previous chapter. Often we write $V(\alpha)$ for $M(\alpha)$.

$$V(\forall x \alpha(x)) = 1 \quad \text{iff} \quad V(\alpha(d)) = 1 \text{ for all } d \in D \qquad (2.7)$$
$$V(\neg \forall x \alpha(x)) = 1 \quad \text{iff} \quad V(\neg \alpha(d)) = 1 \text{ for some } d \in D \;(2.8)$$
$$V(\exists x \alpha(x)) = 1 \quad \text{iff} \quad V(\alpha(d)) = 1 \text{ for some } d \in D \quad (2.9)$$
$$V(\neg \exists x \alpha(x)) = 1 \quad \text{iff} \quad V(\neg \alpha(d)) = 1 \text{ for all } d \in D \quad (2.10)$$

These relations, together with (1.7)–(1.13) of the previous chapter, allow us to define a refutation tree similarly to the way it was done earlier. Let us begin with examples. We again write the formula to be tested at the bottom and proceed bottom-up using mentioned relations.

Example 2.2 Let $\alpha \equiv \neg \forall x P(x) \wedge \forall x (P(x) \wedge Q(x))$.

contradiction

- $\neg P(d), P(d), Q(d)$
- $\neg P(d), P(d) \wedge Q(d)$ by (2.7)
- $\neg P(d), \forall x P(x) \wedge Q(x)$ by (2.8)
- $\neg \forall x P(x), \forall x (P(x) \wedge Q(x))$ by (1.13)
- $\neg \forall x P(x) \wedge \forall x (P(x) \wedge Q(x)) \equiv \alpha$

When d was introduced (at the third line from the bottom) during the bottom-up construction of the tree, it was used as an arbitrary element of the individual domain D. Syntactically it will have to be a new variable, i.e., a variable which does not occur in the bottom line, which in this case is $\neg \forall x P(x), \forall x (P(x) \wedge Q(x))$.

Let us describe the modifications, as done in Section 1.3, for the procedure $\mathcal{P}$ for constructing a refutation tree for a formula (or a list of formulas). The steps $\mathcal{P}^0$, $\mathcal{P}^{\neg\neg}$, $\mathcal{P}^{\neg\vee}$, and $\mathcal{P}^\vee$ are as before. Only step $\mathcal{P}^{val}$ is modified; and steps $\mathcal{P}^\forall$, $\mathcal{P}^\exists$, $\mathcal{P}^{\neg\forall}$, and $\mathcal{P}^{\neg\exists}$ for quantifiers and their negations are added.

Step $\mathcal{P}^\forall$. If Γ contains a formula $\forall x\alpha(x)$, and variable a occurs free in Γ or at a node below Γ, then extend the current node by a new node where Γ is replace by $\Gamma, \alpha(a)$. Do the same if Γ is closed and a is the first variable not in Γ and $\alpha(a)$ is not a member of any node below and including Γ. Notice that Γ is not deleted. Go to the next stage.

Step $\mathcal{P}^\exists$. If Γ contains a formula $\exists x\alpha(x)$ and no formula of the form $\alpha(a)$ for any variable a is a member of any list below and including Γ, then extend the current node by a new node where all formulas in Γ are repeated except that $\exists x\alpha(x)$ is replaced by $\alpha(b)$. The variable b in this case cannot have occurred in Γ or anywhere below it. Go to the next stage.

Steps $\mathcal{P}^{\neg\exists}$, $\mathcal{P}^{\neg\forall}$. These are similar to $\mathcal{P}^\forall$, $\mathcal{P}^\exists$, respectively.

Step $\mathcal{P}^{val}$. If no rule can be applied to a given node, and the list Γ placed at this node is not contradictory, then exit from the procedure with the following verifying model: The domain D of individual variables consists of all variables which occur free in or below Γ. The valuation V is defined by the following relation:

$$\begin{aligned} V(P)(d) &= 1 \text{ if } P(d) \text{ is a member of } \Gamma \\ &= 0 \text{ otherwise.} \end{aligned}$$

A *stipulation*: The steps $\mathcal{P}^\forall$, $\mathcal{P}^{\neg\exists}$ for all possible substitutions of variables (at a given stage) are applied consecutively, i.e., at a row.

This completes the description of procedure $\mathcal{P}$. Consider further examples.

Example 2.3 Let us check the validity of

$$\alpha \equiv \forall x(P(x) \wedge Q(x)) \rightarrow \forall x P(x) \wedge \forall x Q(x).$$

In the following tree, γ is an abbreviation for $\forall x(P(x) \wedge Q(x))$.

$$\text{contradiction} \qquad\qquad \text{contradiction}$$

$$\gamma, P(x_1), Q(x_1), \neg P(x_1) \qquad \gamma, P(x_1), Q(x_1), \neg Q(x_1)$$

$$\gamma, P(x_1) \wedge Q(x_1), \neg P(x_1) \qquad \gamma, P(x_1) \wedge Q(x_1), \neg Q(x_1)$$

$$\gamma, \neg P(x_1) \qquad\qquad \gamma, \neg Q(x_1)$$

$$\gamma, \neg \forall x P(x) \qquad\qquad \gamma, \neg \forall x Q(x)$$

$$\forall x (P(x) \wedge Q(x)), \neg(\forall x P(x) \wedge \forall x Q(x))$$

$$\neg(\forall x(P(x) \wedge Q(x)) \rightarrow \neg(\forall x P(x) \wedge \forall x Q(x)))$$

So the formula α is valid.

Example 2.4 Now consider a formula similar to Example 2.2, where $\wedge$ is replaced by $\vee$:

$$\alpha \equiv (\forall x(P(x) \vee Q(x)) \rightarrow (\forall x P(x) \vee \forall x Q(x))),$$

and let

$$\beta \equiv \neg\alpha,$$
$$\gamma \equiv \forall x(P(x) \vee Q(x)),$$
$$\gamma_1 \equiv P(x_1) \vee Q(x_1)$$

$$\text{contradiction}$$

$$\beta, P(x_1), Q(x_0), \gamma_1, \neg P(x_0), \neg Q(x_1) \quad \beta, Q(x_1), Q(x_0), \gamma_1, \neg P(x_0), \neg Q(x_1)$$

$$\gamma, P(x_1) \vee Q(x_1), Q(x_0), \gamma_1, \neg P(x_0), \neg Q(x_1)$$

$$\text{contradiction} \qquad\qquad |$$

$$\gamma, P(x_0), \gamma_1, \neg P(x_0), \neg Q(x_1) \quad \gamma, Q(x_0), \gamma_1, \neg P(x_0), \neg Q(x_1)$$

$$\gamma, P(x_0) \vee Q(x_0), \gamma_1, \neg P(x_0), \neg Q(x_1)$$

$$\gamma, \neg P(x_0), \neg Q(x_1)$$

$$\gamma, \neg \forall x P(x), \neg \forall x Q(x)$$

$$\forall x(P(x) \vee Q(x)), \neg(\forall x P(x) \vee \forall x Q(x))$$

$$\beta$$

The leftmost upper node (d) is not contradictory and no rule can be applied to it. We have the following model $\langle D, V \rangle$ corresponding to that node.

$$D = \{x_0, x_1\}$$
$$V(P)(x_0) = 0 \qquad V(P)(x_1) = 1$$
$$V(Q)(x_0) = 1 \qquad V(Q)(x_1) = 0$$

We can represent this model graphically as follows:

$$\neg P, Q \qquad\qquad P, \neg Q$$
$$\bullet \qquad\qquad\qquad \bullet$$
$$x_0 \qquad\qquad\qquad x_1$$

The domain consists of all variables occuring free in node d. The values of predicates were computed by the same rule as in the propositional case.

We have $V(\forall x (P(x) \lor Q(x))) = 1$ since for $x = x_0$ the predicate $Q(x)$ (and so $P(x) \lor Q(x)$) is true, and for $x = x_1$, the predicate P is true. On the other hand, we have $V(\neg \forall x P x) = 1$ [take $x = x_0$] and $V(\neg \forall x Q x) = 1$ [take $x = x_1$]. So $V \neg (\forall x P x \lor \forall x Q x) = 1$, and $\neg \alpha$ is satisfied by the assignment V. So α is not valid.

Example 2.5 Consider a formula which is in fact equivalent to the formula in Example 2.3. Here a trick will be used to construct a finite model.

$$\alpha \equiv \forall x \exists y \, ((P(x) \lor Q(x)) \land \neg Q(y)) \rightarrow \forall x P(x).$$

Let $\neg \alpha \equiv \forall x \exists y \; \beta(x, y) \land \neg \forall x \; P(x)$ where $\beta(x, y) \equiv ((P(x) \lor Q(x)) \land \neg Q(y))$ or $\gamma \land \neg \forall x \; P(x)$.

$$\begin{array}{cc}
[\text{ take } x_2 = x_1] & \text{contradiction} \\
\gamma, P(x_1), \neg Q(x_2), Q(x_0), \neg Q(x_1), \neg P(x_0) & \gamma, Q(x_1), \neg Q(x_1),
\end{array}$$

$$\gamma, P(x_1) \lor Q(x_1), \neg Q(x_2), Q(x_0), \neg Q(x_1), \neg P(x_0)$$
$$\gamma, \beta(x_1, x_2), [\equiv (P(x_1) \land Q(x_1)) \land \neg Q(x_2)], Q(x_0), \neg Q(x_1), \neg P(x_1)$$
$$\gamma, \exists y \beta(x_1, y), Q(x_0), \neg Q(x_1), \neg P(x_0)$$

contradiction

$$\frac{\gamma, P(x_0), \neg Q(x_1), \neg P(x_0) \quad \gamma, Q(x_0), \neg Q(x_1), \neg P(x_0)}{\gamma, P(x_0) \vee Q(x_0), \neg Q(x_1), \neg P(x_0)}$$

$\gamma, \beta(x_0, x_1), \neg P(x_0)$

[where $\beta(x_0, x_1) \equiv (P(x_0) \vee Q(x_0)) \wedge \neg Q(x_1)$]

$\gamma, \exists y \beta(x_0, y), \neg P(x_0)$

$\gamma, \neg P(x_0)$

$\gamma, \neg \forall x P(x)$

$\forall x \exists y \beta(x, y) \wedge \neg \forall x P(x)$

To the only remaining node no rule is applicable, and we get the same model as in Example 2.2.

For a more difficult example, the reader might try to find a falsifying model for the formula

$$\forall x \exists y ((P(x) \wedge Q(y)) \vee (Q(x) \wedge P(y)))$$
$$\rightarrow \exists y \forall x ((P(x) \wedge (Q(y)) \vee (Q(x) \wedge P(y))))$$

2.4 Termination Proof for Modal-like Formulas

The rules stated above are sufficient to prove any valid formula, but are not sufficient in general to obtain finite models (since $\mathcal{P}$ will not always terminate), although any non-contradictory monadic predicate formula does in fact possess a finite model. Still the refutation procedure $\mathcal{P}$ terminates on the subclass of formulas which will be sufficient for our aims.

Definition. A formula α of monadic predicate logic is *modal-like* if any subformula of α contains free at most one individual variable, and no free variable of α is in the scope of any quantifier.

Example 2.6 The following formula

$$\alpha \equiv (P(y) \vee Q(y)) \wedge \forall x (\neg P(x) \wedge \neg Q(x) \wedge$$
$$\forall x \exists x (P(x) \vee Q(x)))$$

is modal-like, but formulas

$$\beta \equiv P(x) \vee Q(y), \qquad\qquad (2.11)$$
$$\gamma \equiv P(y) \vee \forall x (P(x) \vee Q(y))$$

are not since each of them contains subformula $P(x) \vee Q(y)$.

The term *modal-like* is derived from the following translation *mod* of modal-like formulas into modal formulas: erase all arguments in the predicates and replace $\forall$, $\exists$ by $\square$, $\Diamond$, respectively (changing capital predicate letters into lower case propositional letters).

Example 2.6 (continued) The translation yields

$$\alpha^{mod} \equiv (p \vee q) \wedge \square(\neg p \wedge \neg q \wedge \square\Diamond(p \vee q))$$

and all information present in α is completely preserved. On the other hand

$$\beta^{mod} \equiv p \vee q,$$

$$\gamma^{mod} \equiv p \vee \square(p \vee q)$$

and the difference between the arguments of P and Q in β and γ disappears (hence information is lost).

Let us analyze the structure of arbitrary formulas which can occur in refutation trees. It turns out that such formulas are subformulas of the final formula up to substitution of individual variables and prefixing negation.

The sign, with or without negation, of a subformula in the refutation tree for α is determined by the sign of its ancestor in α, which is defined as follows. Let α be a formula containing connectives $\neg, \vee, \wedge, \forall, \exists$. The sign of a subformula (occurrence) β in a formula α is $+$ (i.e., β occurs positively) if β is in the scope of an even number of occurrences of negation, and the sign on β is $-$ (β occurs negatively) otherwise.

Example 2.7

$$
\begin{array}{cccccccccccc}
+ & ++ & -+ & - & + & + & + & + & - & - \\
\alpha \equiv p \wedge (& \neg\, q & \vee\, \neg\, (p \wedge \neg\, q)\,) & \wedge & \forall x(R(x) & \wedge & \neg\forall y & R(y)) \\
& + & \underline{\quad + \quad} & & & & &
\end{array}
$$

The signs above are as follows: the α, the first p, first $\neg q$, $\neg(p \wedge \neg q), R(x)$, and the second q occur positively, while the first q, $(p \wedge \neg q)$, the second p, $R(y)$, and the second $\neg q$ occur negatively. If the formula α also contains the implication sign $\rightarrow$, then the premiss of the $\rightarrow$ is counted as a negation: β occurs positively if it is in the scope of an even number of negations and premisses of implication.

This definition agrees with the translation of $\alpha \rightarrow \beta$ as $\neg\alpha \vee \beta$.

Example 2.8

$$p \; \wedge \; \neg \; (q \; \rightarrow \; \neg \; (p \; \rightarrow \; q \;))$$
$$+ \; + \; + \quad + \; - \; - \quad \; - \; + \; +$$
$$p \; \wedge \; \neg \; (\neg q \; \vee \; \neg \; (\neg p \; \vee \; q \;))$$

The sign of a subformula in a list Γ of formulas is the same as in the corresponding element of Γ.

Theorem 2.1 *Let Γ be a list of formulas, let T be a refutation tree for Γ, and let α be a formula appearing in T as a member formula at some node. Then*

$$\alpha \equiv \pm\beta_{x_1...x_n}[y_1, \ldots, y_n] \tag{2.12}$$

where $\pm$ means possible presence of negation; $\overline{x} \equiv \{x_1, \ldots, x_n\}$ and $\overline{y} \equiv \{y_1, \ldots, y_n\}$ are lists of individual variables; and β is a subformula of Γ, $+\beta$ means β, and $-\beta$ means $\neg\beta$, and the sign is the same as the sign of β in Γ.

Proof. The proof is by induction on the level l of the tree.

The base case $l = 0$ refers to the bottom node, where each member formula is obviously a positive subformula of Γ.

The induction step is based on the fact that each rule possesses the required property. More precisely, the member formula in the premise of the rule either is directly taken from the conclusion, or is introduced by the current rule. In the first instance, relation (2.13) is preserved in the conclusion, by inductive assumption. In the second instance, α is replaced by either α_1 or α_1, α_2 (including the case for rules $\mathcal{P}^\vee$, $\mathcal{P}^{\neg\exists}$ when α is preserved in the conclusion, so that $\alpha_2 \equiv \alpha$). Consider the possible cases in turn, abbreviating (2.13) in the shorter form

$$\alpha \equiv \pm\beta^* \tag{2.13}$$

Case 1. Step $\mathcal{P}^{\vee}$. Then $\alpha \equiv \alpha_1 \vee \alpha_2$, and the step is

$$\frac{\alpha_1, \alpha_2, \Delta}{\alpha \vee \alpha_2, \Delta}$$

and the sign in (2.14) should be $+$, since α does not begin with $\neg$. So $\beta \equiv \beta_1 \vee \beta_2$, and we put $\alpha_i \equiv \beta_i^*$, as required.

Case 2. Step $\mathcal{P}^{\neg\neg}$. Then $\alpha \equiv \neg\neg\alpha_1$.

$$\frac{\alpha_1, \Delta}{\neg\neg\alpha_1, \Delta}$$

If the sign in (2.14) is $+$, i.e., $\neg\neg\alpha_1 \equiv \alpha \equiv \beta^*$, then β should begin with $\neg\neg$, i.e., $\beta \equiv \neg\neg\beta_1$ and we have $\alpha_1 \equiv \beta_1^*$. Otherwise $\neg\neg\alpha_1 \equiv \neg\beta$, so $\beta \equiv \neg\beta_1$ and again $\alpha_1 \equiv \beta_1^*$.

Case 3. Step $\mathcal{P}^{\neg\vee}$. Then $\alpha \equiv \neg(\alpha_1 \vee \alpha_2)$. If the sign in (2.14) is $+$, we have $\beta \equiv \neg(\beta_1 \vee \beta_2)$ and $\alpha_i \equiv \neg\beta_i^*$ as required. If the sign in (2.14) is minus, we have $\alpha \equiv \neg\beta^*$ which implies $\beta \equiv \beta_1 \vee \beta_2$, and $\alpha_i \equiv \beta_i^*$.

Case 4. Step $\mathcal{P}^{\exists}$. Then $\alpha \equiv \exists z\gamma$ and the formula α_1 to be added to the list at this step is $\gamma_z[u]$. Here the sign in (2.14) is $+$, i.e., $\exists x\gamma \equiv \alpha \equiv \beta^*$, so that $\beta \equiv \exists x\delta$ and $\gamma \equiv \delta^*$. So $\alpha_1 \equiv \gamma_x[u] \equiv (\delta_{x_1,\ldots,x_n}[y_1,\ldots,y_n])_z[u] \equiv \delta_{x_1,\ldots,x_n z}[y_1,\ldots,y_n,u]$, as required.

Case 5. Step $\mathcal{P}^{\neg\exists}$ Treated as the combination of cases 3 and 4. This concludes the proof of Theorem 2.1. $\square$

We assume now that no formula is analyzed twice in one and the same branch of procedure $\mathcal{P}$, although this is in fact not necessary for the truth of the next theorem.

Theorem 2.2 *Procedure $\mathcal{P}$ terminates for any modal-like formula.*

Proof. For any modal-like formula α, a number N_α will be determined such that the length (i.e., the number of levels) of any branch in the refutation tree is bounded by N_α. We begin by proving the following lemma.

Lemma 2.3 *In any refutation tree*

    (a) No step $\mathcal{P}^{\exists}$ or $\mathcal{P}^{\neg\forall}$ is applied to one and the same formula in the same branch; and

(b) *The number of times that the rule $\mathcal{P}^\forall$ or $\mathcal{P}^{\neg\exists}$ is applied to the same formula in the same branch is bounded by*

$$1 + \text{ the number of steps in which } \mathcal{P}^\exists, \mathcal{P}^{\neg\forall}$$
$$\text{introduce new variables in that branch} \qquad (2.14)$$

Proof. (a) Each step of the type $\mathcal{P}^\exists$, $\mathcal{P}^{\neg\forall}$ applied to $\pm Qx\alpha$ (for respective quantifier Q) replaces it by $\pm\alpha_x[u]$, with the new variable u. A second such step applied to the same formula in the same branch is prohibited by the restriction in the description of the rule.

(b) To apply a step of the type $\mathcal{P}^\forall$, $\mathcal{P}^{\neg\exists}$ to the same formula again, it is necessary to have a new free variable. Such a variable can be introduced the very first time by $\mathcal{P}^\forall$ or $\mathcal{P}^{\neg\exists}$ applied to a list consisting of closed formulas, which accounts for the first 1 in formula (2.15), and then once for each new variable that steps of type $\mathcal{P}^\exists$, $\mathcal{P}^{\neg\forall}$ introduce, which accounts for the second summand. This concludes the proof of the lemma. $\qquad\square$

Proof of Theorem 2.2 continued. Let α be a modal-like formula containing q occurences of quantifiers and S occurences of propositional connectives. Let

$$N_\alpha \equiv q + (q+1)(S+q).$$

Since α is modal-like, note that for any subformula $Qz\beta$ of α beginning with a quantifier Q, either β does not contain free variables except z, or β does not contain z free (i.e., Qz is redundant). By Theorem 2.1, any formula analyzed in a quantifier step of procedure $\mathcal{P}$ has the form $\pm Qz\beta^*_z$ where $Qz\beta$ is a subformula of the original formula α, and $*$ denotes some substitution for free variables in $Qz\beta$.

New variables can be introduced at such a quantifier step only if z occurs free in β, or else the substitution is empty. So any subformula $Qz\beta$ of α generates at most one such step, one for the sign $+$, or $-$ in $\pm Qz\beta$, and the estimate (2.15) leads to the multiplier $(q+1)$ in N_α:

in one branch there are at most q steps of type
$\mathcal{P}^\exists$ or $\mathcal{P}^{\neg\forall}$ introducing new variables. $\qquad (2.15)$

By Lemma 2.3(b), there are no more than $(q+1)$ steps $\mathcal{P}^\forall$, $\mathcal{P}^{\neg\exists}$ in one branch analyzing the same formula $\pm Qz\beta$, and each subformula of α beginning with a quantifier generates at most one such step (for a plus or minus sign). So, (2.16) implies that

$$\text{the total number of the steps } \mathcal{P}^\forall, \ \mathcal{P}^{\neg\exists} \text{ in}$$
$$\text{one branch is at most } q(q+1). \tag{2.16}$$

Let us then count the total number of propositional steps in one branch of the refutation tree. By Theorem 2.1 every formula analyzed at any step of the tree T has the form $\pm\beta^* \equiv \pm\beta_{\bar{x}}[\bar{y}]$ where β is a subformula of α and $\bar{x}$, $\bar{y}$ are sets of variables, $x_1, \ldots, x_n$ and $y_1, \ldots, y_n$, respectively. Since α is modal-like, no more than one variable can occur free in β, so $\bar{x}$ can be assumed to be a singleton, i.e., a list consisting of one variable only. By (2.16) the total number of possible values for $\bar{y}$ is bounded by $q+1$. So, since any propositional connective is analyzed only once,

$$\text{there are no more than } (q+1)S \text{ propositional}$$
$$\text{steps in one branch} \tag{2.17}$$

where S is the number of propositional connectives in α.

The total number of steps therefore is a sum corresponding to steps of type $(\mathcal{P}^\exists, \mathcal{P}^{\neg\forall})$, $(\mathcal{P}^\forall, \mathcal{P}^{\neg\exists})$, as well as propositional steps, and so is bounded by $q + q(q+1)q + (q+1)S = q + (q+1)(S+q) = N_\alpha$, as was to be proved. $\qquad\square$

2.5   Soundness and Completeness

We next prove the soundness and completeness of the refutation procedure $\mathcal{P}$. To do this, let us first summarize some of its features, not restricting ourselves to modal-like formulas. Recall that a *branch* of any refutation tree is a sequence of nodes, beginning with the bottom node, and containing at each node also one of its immediate predecessors (if the node in question has any, i.e., is not a leaf of the tree). The branch in a refutation tree is *closed* if it contains a contradiction α, $\neg\alpha$. The next lemma summarizes some properties of branches in refutation trees.

Lemma 2.4 *Let B be a finished but non-closed branch in the refutation tree. Then*

$(\vee)$ *if $(\alpha \vee \beta) \in B$ then $\alpha \in B$ or $\beta \in B$;*

$(\neg\vee)$ *if $\neg(\alpha \vee \beta) \in B$ then $\neg\alpha \in B$ and $\neg\beta \in B$;*

$(\neg\neg)$ *if $\neg\neg\alpha \in \beta$ then $\alpha \in \beta$;*

$(\exists)$ *if $\exists x\alpha \in \beta$ then $\alpha_x[u] \in \beta$*
          for some variable u which occurs free in B;

$(\forall)$ *if $\forall x\alpha \in \beta$ then $\alpha_x[u] \in B$*
          for all variables u which occur free in B;

$(\neg\exists)$ *if $\neg\exists x\alpha \in B$ then $\neg\alpha_x[u] \in B$*
          for all variables u which occur free in B;

$(\neg\forall)$ *if $\neg\forall x\alpha \in \beta$ then $\neg\alpha_x[u] \in B$*
          for some variable u which occurs free in B.

Proof. If a formula having one of the forms above is a member of B, Theorem 2.4 says what happens when it is analyzed at some stage of the procedure. All possible values u for variable x in $(\forall)$ and $(\neg\exists)$ are eventually substituted. The only thing to check is what happens when some step is not made for a formula occuring in B due to the restrictions. In the propositional cases $(\vee)$, $(\neg\vee)$, and $(\neg\neg)$, this can happen since the step would be repeated analysis of the same formula. But then it was already analyzed, so necessary subformulas are already present in B. For case $(\exists)$, analysis is not done exactly in the case when the $(\exists)$-clause is satisfied, i.e., $\alpha[u] \in B$ for some u. The same is true for the remaining quantifier cases. This completes the proof. $\square$

Let us reformulate refutatation procedure $\mathcal{P}$ into a Gentzen-type system G for inferring contradictions, exactly as it was done in Chapter 1 for the propositional case.

Derivable objects are again sequents, i.e., lists of predicate formulas, considered up to permutation. Axioms and propositional rules are as before, and four predicate rules are added, namely, the rules for $\exists$, $\forall$, $\neg\exists$, $\neg\forall$.

Deductive System G (predicate version).

○ Axioms: $\Gamma, \alpha, \neg\alpha$

○ Inference rules: $(\vee), (\neg\vee), (\neg\neg)$ as above; and

$$\frac{\Gamma, \alpha_x[b]}{\Gamma, \exists x\alpha}\,(\exists) \qquad\qquad \frac{\Gamma, \forall x\alpha, \alpha_x[u]}{\Gamma, \forall x\alpha}\,(\forall)$$

$$\frac{\Gamma, \neg\alpha_x[b]}{\Gamma, \neg\forall x\alpha}\,(\neg\forall) \qquad\qquad \frac{\Gamma, \neg\exists x\alpha, \neg\alpha[u]}{\Gamma, \neg\exists x\alpha}\,(\neg\exists)$$

where variable b in the rules $(\exists), (\neg\forall)$ should be new, i.e., not occur free in the conclusion. Recall that formula lists above the line are *premisses* of the rule.

Lemma 2.5 *All the rules of system G are sound and invertible: that is, for each of these rules, the conclusion is true in a model $M \equiv \langle D, V \rangle$ under some substitution S of objects of D for free variables iff at least one of the premisses of the rule is true under some extension of the substitution S to new variables.*

Proof. For the propositional rules $\mathcal{P}^{\neg\neg}$, $\mathcal{P}^{\vee}$, and $\mathcal{P}^{\neg\vee}$ there are no new variables betweeen the premisses and the conclusion, and our Lemma is reduced to the first Lemma of Section 1.4. Each of the predicate rules $(\forall), (\neg\forall), (\exists), (\neg\exists)$ can be written in the form Π'/Π where Π' is a premiss, and Π is the conclusion. For the rules $(\forall)$, and $(\neg\exists)$, we have $M(\Pi'^*) = M(\Pi^*)$ for any model $M = \langle D, V \rangle$ and any substitution $*$ of objects in D for free variables since

$$M((\forall x\alpha)^* \wedge (\alpha[u])^*) = M((\forall x\alpha)^*)$$

$$M((\neg\exists x\alpha)^* \wedge (\neg\alpha[u])^*) = M((\neg\exists x\alpha)^*)$$

For the rules $(\exists)$, and $(\neg\forall)$ use the following relations:

$$M((\exists x\alpha)^*) = \min_{d \in D} M(\alpha^*{}_x[d]);$$

$$M((\neg\forall x\alpha)^*) = \min_{d \in D} M(\neg\alpha^*{}_x[d])$$

This concludes the proof.   □

Theorem 2.6 *Refutation procedure $\mathcal{P}$ is sound for all formulas. That is, if a model M is produced by $\mathcal{P}$ when applied to formula α, then $M(\alpha) = 1$; and if the result of $\mathcal{P}$ for α is negative, then α is contradictory.*

Proof. The proof for the negative case is exactly the same as for the propositional calculus. For a positive result, the proof is more complicated than for Theorem 1.2(a). Recall that the model $\langle D, V \rangle$ produced by procedure $\mathcal{P}$ from a terminal node is determined by formulas and variables occuring in branch $\mathcal{B}$ of the refutation tree leading to this node.

$$D = \{v : v \text{ is an individual variable occuring free in } \mathcal{B}\} \quad (2.18)$$

$$V(\mathcal{P})(d) = 1 \text{ iff } \mathcal{P}(d) \text{ occurs in } \mathcal{B} \text{ as a member formula} \quad (2.19)$$

We shall prove that (2.19) extends to arbitrary formulas by showing that:

$$\text{If } \gamma \in \mathcal{B}, \text{ then } V(\gamma) = 1 \quad (2.20)$$

(2.20) is proved by induction on the construction of γ, i.e., on the number l of logical connectives in γ. The base case $l = 0$ for (2.20), i.e., the case when $\gamma \equiv \mathcal{P}(d)$, is just (2.19). If $\neg \mathcal{P}(a)$ occurs in $\mathcal{B}$, then $\mathcal{P}(a)$ does not occur there, since $\mathcal{B}$ is not contradictory. This implies (2.20) for $\gamma \equiv \neg \mathcal{P}(a)$ in the induction step. The rest of the induction step is proved by cases, depending on the main connective of γ, corresponding to the different cases in Lemma 2.4.

Cases $(\vee)$, $(\neg \vee)$. Let $\gamma \equiv \alpha \vee \beta \in \mathcal{B}$. Then $\alpha \in \mathcal{B}$ or $\beta \in \mathcal{B}$ by Lemma 2.4, so $V(\alpha) = 1$ or $V(\beta) = 1$ by the induction hypothesis; that is $V(\gamma) \equiv V(\alpha \vee \beta) = 1$ as required. Now let $\neg(\alpha \vee \beta) \in \mathcal{B}$. Then $\neg \alpha, \neg \beta \in \mathcal{B}$, hence $V(\neg \alpha) = V(\neg \beta) = 0$, and so $V((\alpha \vee \beta)) = 0 = V(\gamma)$, as required.

Case $(\neg \neg)$. Let $\gamma \equiv \neg \neg \alpha \in \mathcal{B}$. Then $\alpha \in \mathcal{B}$ and $V(\alpha) = 1 = V(\neg \neg \alpha) = V(\gamma)$ as required.

Cases $(\exists)$, $(\neg \exists)$. Let $\gamma \equiv \exists x \alpha \in \mathcal{B}$. Then $\alpha_x[u] \in \mathcal{B}$ for some variable u which occurs free in $\mathcal{B}$ (and so belongs to the individual domain D of the model M). By the induction hypothesis $V(\alpha[u]) = 1$, so $V(\exists x \alpha) = 1 = V(\gamma)$ as required. If $\neg \exists x \alpha \in \mathcal{B}$, then $\neg \alpha_x[u] \in \mathcal{B}$ for all variables $u \in D$, and hence $V(\alpha[u]) = 0$

for all such u by the induction hypothesis. This implies that $V(\exists x\alpha) = 0 = V(\gamma)$, as required.

Cases $(\forall)$, $(\neg\forall)$. are treated similarly to cases $(\exists)$ and $(\neg\exists)$, and so (2.21) is proved.

Since α is present at $\mathcal{B}$, namely at the bottom node, we have $V(\alpha) = 1$, which completes the proof. $\Box$

Theorem 2.7 *The refutation procedure $\mathcal{P}$ is complete. That is, let α be a modal-like formula, and let $d_1, \ldots, d_n$ (or shorter $\bar{d}$) be individuals of some model $M = \langle \mathcal{D}, V \rangle$. If $M(\alpha_{\bar{x}}[\bar{d}]) = 1$, then there is a non-contradictory branch $\mathcal{B}$ in the refutation tree such that the corresponding model M' satisfies α, i.e., $M'(\alpha_{\bar{x}}[\bar{d}']) = 1$ under the suitable correspondence between the individual domains of M and M'.*

Proof. The proof is essentially the same as for Theorem 1.2(b). To find the required branch $\mathcal{B}$, proceed bottom-up through the given refutation tree. At the nodes where the refutation tree branches, choose the predecessor in the tree which is (or rather contains the list) true in the model M. The uppermost node of the branch $\mathcal{B}$ is also true under that model. In the course of this bottom-up process, one should accumulate substitutions of objects from D for new individual variables which appear in $\mathcal{B}$. This always happens in the steps $\mathcal{P}^{\exists}$, $\mathcal{P}^{\neg\forall}$ and can happen at $\mathcal{P}^{\forall}$, $\mathcal{P}^{\neg\exists}$. Consider each of these cases in turn.

Case $\mathcal{P}^{\exists}$. We have $V(\Gamma^*, \exists x\alpha^*) = 1$, where $*$ means substitution of objects from D for free variables. Then there is $d \in D$ such that $V(\alpha^*{}_x[d]) = 1$. Adjoin substitution $x := d$ to the $*$-substitution, and we have $V(\Gamma^*, \alpha^*) = 1$, as required.

Case $\mathcal{P}^{\forall}$. We have $V(\Gamma^*, \forall x\alpha^*) = 1$, so

$$V(\alpha^*{}_x[d]) = 1 \tag{2.21}$$

for all $d \in D$ where $*$ is a substitution for free variables different from x. We have to prove that

$$V((\alpha_x[u])^{**}) = 1 \tag{2.22}$$

for all variables u occuring free in $\mathcal{B}$, where $**$ includes substitution for u. If u is among the variables involved in $*$, then take

$d = u^*$ in (2.22). If u is a new variable, take d to be arbitrary in (2.22), and (2.23) is proved.

Cases $\mathcal{P}^{\neg\exists}$, $\mathcal{P}^{\neg\forall}$. These are treated similarly to $\mathcal{P}^\forall$, $\mathcal{P}^\exists$ respectively, and the proof is concluded.   □

Let us reformulate Theorem 2.6 as the completeness result for system G.

Corollary 2.8 *The deductive system* G *is sound and complete for modal-like formulas α. That is, any formula α is valid just in case $\neg\alpha$ is derivable in* G.

Proof. Suppose first that $\neg\alpha$ is derivable in G. Then, since axioms are obviously false in any model and all rules preserve falsity by Lemma 2.5, the formula $\neg\alpha$ is also false in any model. This means that α is valid, as required.

Let now α be valid. Then $\neg\alpha$ is contradictory, i.e., false in any model. So the procedure $\mathcal{P}$ applied to $\neg\alpha$ cannot give a positive answer, that is, the answer is negative. But then the refutation tree is a required derivation of $\neg\alpha$, which concludes the proof.   □

In fact, the system G is sound and complete for all (not only modal-like) predicate formulas, and this is easily proved by an extension of the methods presented here.

3

The System S5

3.1 The Syntax of Modal Logic

The system S5 will be our first system of propositional modal logic. All of the systems will have one and the same set of well-formed formulas in the *language of the propositional modal logic*, which is obtained by adding modal connectives $\Box$ (it is necessary) and $\Diamond$ (it is possible) to the language of the classical propositional logic.

More precisely, *formulas* of modal propositional logic (or modal propositional formulas) are constructed from the propositional letters by means of $\neg$, $\vee$, $\Box$, $\Diamond$.

The following are examples of modal propositional formulas:

$$p, \ \neg p \vee q, \ p \vee \Box q, \ \neg \Box \Diamond p \vee \Diamond \Box p, \ \Box(\neg \Box p \vee p)$$

The other boolean connectives ($\wedge$, $\rightarrow$, $\leftrightarrow$), are treated as abbreviations, as in previous chapters.

3.2 Semantics: Possible Worlds and Valuations

We are going to define a semantics of possible worlds for modal formulas where the relative possibility of worlds is understood in the most liberal way: all worlds are possible. So the possibility relation between worlds (not to be confused with the possibility operator ($\Diamond$) for sentences) can be ignored altogether. Let L denote the set of all propositional letters.

An S5-*model* is a pair $\langle W, V \rangle$ where W is a nonempty set, and V is a function from the set $L \times W$ into truth-values

$$V : L \times W \to \{0, 1\} \qquad (3.1)$$

The relation (3.1) means that $V(p, w) \in \{0, 1\}$, i.e., it is a truth-value for any propositional letter p and element $w \in W$. The set W is called a *frame* of the model $\langle W, V \rangle$, and its elements are *worlds*. The function V is a *valuation* of propositional letters in the worlds of the model $\langle W, V \rangle$.

For fixed world $w \in W$ the values $V(p, w)$ for various letters p show the *truth-values of p in the world w*. Some of these letters are true, and some are false, so the totality of these truth-values for a fixed w and different letters

$$V(p_1, w), V(p_2, w), \ldots V(q, w), \ldots \qquad (3.2)$$

describes the state of affairs in the world w: what is true and what is false. In this sense, elements $w \in W$ can intuitively be called worlds. Strictly speaking, the sequence (3.2) describes the truth-values not for all sentences, but only for atoms, from which other sentences are constructed. Now, however, the valuation will be extended to complex formulas, α, so that $V(\alpha, w) = 1$, using the standard interpretation of boolean connectives $\neg, \vee$ and the understanding of $\Box \alpha$ as "α is true in all worlds," and of $\Diamond \alpha$ as "α is true in some world." We define:

$$V(\neg \alpha, w) = 1 \quad \text{iff} \quad V(\alpha, w) = 0; \qquad (3.3)$$
$$V(\alpha \vee \beta, w) = 1 \quad \text{iff} \quad (V(\alpha, w) = 1 \text{ or } V(\beta, w) = 1);$$
$$V(\Box \alpha, w) = 1 \quad \text{iff} \quad V(\alpha, w_1) = 1 \text{ for all } w_1 \in W;$$
$$V(\Diamond \alpha, w) = 1 \quad \text{iff} \quad V(\alpha, w_1) = 1 \text{ for some } w_1 \in W.$$

Let us consider some examples:

Example 3.1 $V(p \vee \neg p, w) = 1$ for any world w and valuation V. Indeed,

$$
\begin{aligned}
V(p \vee \neg p, w) = 1 \quad &\text{iff} \quad (V(p, w) = 1 \text{ or } V(\neg p, w) = 1) \\
&\text{iff} \quad (V(p, w) = 1 \text{ or } V(p, w) = 0)
\end{aligned}
$$

but the latter condition is always satisfied, since $V(p, w) \in \{0, 1\}$.

Example 3.2 Let our model have two worlds, I and II, and p is false in the first world, and is true in the second world. How about $\Box p \vee \Box \neg p$? We have $V(p, \mathrm{I}) = 0$, $V(p, \mathrm{II}) = 1$. Now

$$V(\Box p \vee \Box \neg p, \mathrm{I}) = 1 \quad \leftrightarrow \quad (V(\Box p, \mathrm{I}) = 1 \quad \text{or} \quad V(\Box \neg p, \mathrm{I}) = 1)$$
$$\leftrightarrow \quad (\forall w_1 V(p, w_1) = 1 \quad \text{or} \quad \forall w_1 V(\neg p, w)) = 1)$$
$$\leftrightarrow \quad (\forall w_1 V(p, w_1) = 1 \quad \text{or} \quad \forall w_1 V(p, w_1) = 0)$$

In the last sentence, the first disjunct is false when $w_1 =\mathrm{I}$, and the second disjunct is false when $w_1 =\mathrm{II}$. So we see that the sentence $\Box p \vee \Box \neg p$ is false in world I of our model, which we draw as follows:

$$
\begin{array}{ccc}
\mathrm{II} & \bullet & p \\
& \Big\uparrow & \\
\mathrm{I} & \bullet & \neg p
\end{array}
$$

World II is placed above I, but this does not have any significance in this situation since the set of worlds is completely homogeneous.

Example 3.3 $\alpha \equiv \Box p \to p \equiv \neg \Box p \vee p$

$$V(\alpha, w) = 1 \quad \leftrightarrow \quad (V(\Box p, w) = 0 \vee V(p, w) = 1)$$
$$\leftrightarrow \quad \neg(\forall w_1 V(p, w_1) = 1) \vee V(p, w) = 1$$
$$\leftrightarrow \quad \forall w_1 V(p, w_1) = 1 \to V(p, w) = 1$$

which is obvious: instantiate w_1 by w.

So $\Box p \to p$ is true in any world w.

Formula α is *valid* in a model $\langle W, V \rangle$ if α is true in every world $w \in W : V(\alpha, w) = 1$.

Formula α is S5-valid, if it is true in every model $\langle W, V \rangle$.

Let us increase our stock of valid formulas. Note that the propositional clauses (i.e., $\neg, \vee$) in the definition of truth in a world (3.3) are exactly the same as in classical propositional logic. Furthermore, the modal clauses (i.e., $\Box, \Diamond$) remind us of the clauses for quantifiers $\forall, \exists$ in the truth definition for the classical predicate logic. We shall see that this is not accidental.

3.3   Comparison with Monadic Predicate Logic

For any modal formula α and individual variable x, define monadic predicate formula α^x in the following way. Replace all modal signs $\Box, \Diamond$ by quantifiers $\forall x, \exists x$ respectively, and all propositional letters p by $P(x)$, i.e., capitalize propositional letters and add argument x, one and the same individual variable for all formulas.

Let us compare the translations α^{mod} of modal-like formulas into modal predicate formulas x (introduced in the previous chapter), and the translation α^x.

We claim they are inversions of each other.

Lemma 3.1

(a) $(\alpha^x)^{mod} \equiv \alpha$ *for all modal formulas α*   (3.4)

(b) $(\alpha^{mod})^x \equiv \alpha$ *for closed modal-like, monadic* (3.5)
         predicate formulas α up to re-
         naming of bound variables into
         x;

(c) $(\alpha^{mod})^x \equiv \alpha_b[x]$ *for all modal-like formulas α*   (3.6)
         up to renaming bound vari-
         ables into x, where b is the
         (only) variable which can oc-
         cur free in α.

Proof. Induction base. α is atomic.

Relation (3.5) is a particular case of (3.6). Relations (3.4) and (3.6) are easily proved by induction on the construction of α.

Ad(3.4) We have $\alpha \equiv p$. Then $\alpha^x \equiv P(x)$, and $(\alpha^x)^{mod}$ is
    again p as required.

Ad(3.6) $\alpha \equiv P(b)$. Then $\alpha^{mod} \equiv p$, and $(\alpha^{mod})^x \equiv P(x) \equiv$
    $P(b)_b[x]$ as required.

Induction step. Consider possible cases.

Case $\vee$. Here $\alpha \equiv (\beta \vee \gamma)$. Then by induction hypothesis we have (3.4), (3.6) for β and γ instead of α. Let us obtain them for α.

Ad(3.4) We have

$$\alpha^x \equiv \beta^x \vee \gamma^x, (\alpha^x)^{mod} \equiv (\beta^x)^{mod} \vee (\gamma^x)^{mod}.$$

By the induction hypothesis this implies

$$(\alpha^x)^{mod} \equiv \beta \vee \gamma \equiv \alpha.$$

Ad(3.6) We have $\alpha^{mod} \equiv \beta^{mod} \vee \gamma^{mod}$ for modal-like α, so

$$(\alpha^{mod})^x \equiv (\beta^{mod})^x \vee (\gamma^{mod})^x \equiv \beta_b[x] \vee \gamma_b[x]$$

by induction hypothesis applied to β and γ. The latter formula is $\alpha_b[x]$ as required.

Case $\neg$. Similar to the previous.

Case $\forall$. Here $\alpha \equiv \forall y \beta$, and y is the only variable free in β. We have (3.6) for β instead of α, and have to prove only (3.6) for α, since (3.4) is not applicable. We have

$$\alpha^{mod} \equiv \Box \beta^{mod}, (\alpha^{mod})^x \equiv \forall x (\beta^{mod})^x \equiv \forall x \beta_y[x],$$

the latter step by the induction hypothesis. But $\forall x \beta_y[x]$ is exactly the result of renaming bound variable y into x in the formula $\alpha \equiv \forall y \beta$ as was to be proved.

Case $\Box$. Here $\alpha \equiv \Box \beta$, and we have to prove only (3.4). We have $\alpha^x \equiv \forall x \beta^x$, so

$$(\alpha^x)^{mod} \equiv \Box (\beta^x)^{mod} \equiv \Box \beta$$

by induction hypothesis.

The cases $\exists$ and $\Diamond$ are treated similarly. This concludes the proof. $\qquad \Box$

Lemma 3.1 establishes a syntactical connection between α^{mod}, α^x, and α. We shall now show that there is also a very close semantic connection: modal formula α has the same meaning as α^{mod}. To state this connection in full generality we need a corresponding connection between models, which turns out to be very simple. We consider any model $\langle D, V \rangle$ for predicate modal logic (with individual domain D and valuation V) to be at the same time the modal model with the set of worlds D and valuation V. More precisely, we put in the modal case

$$\tilde{V}(p, w) = V(P)(w) \text{ for any } w \in D.$$

We denote modal model $\langle D, V \rangle$ corresponding to $M \equiv \langle D, V \rangle$ by $\widetilde{M}$.

So the objects which are assigned values by the predicate model $M \equiv \langle D, V \rangle$ are closed formulas of predicate logic, as well as objects $\alpha[w_1, \ldots, w_n]$ where α is a formula and $w_1, \ldots, w_n \in D$, i.e., $w_1, \ldots, w_n$ are worlds in the corresponding modal model $\widetilde{M}$. On the other hand, valuation $\tilde{V}$ of the model $\widetilde{M}$ assigns truth values to pairs (α, w) where α is a modal formula, and $w \in D$. These values are related as follows.

Theorem 3.2 *For any predicate model $M \equiv \langle D, V \rangle$ and corresponding modal model $\widetilde{M} = \langle D, \tilde{V} \rangle$*

$$(a) \quad \tilde{V}(\alpha, w) = V(\alpha_x^x[w]) \tag{3.4}$$

for any modal formula α and any $w \in D$;

$$(b) \quad \tilde{V}(\alpha^{mod}, w) = V(\alpha_b[w]) \tag{3.5}$$

for any modal-like monadic predicate formula α, the (only) variable b which occurs free in α, and any $w \in D$.

Note. For closed formulas these relations are simpler:

$$\tilde{V}(\alpha, w) = V(\alpha^x) \text{ if } \alpha \text{ is modalized.}$$

$$\tilde{V}(\alpha^{mod}, w) = V(\alpha) \quad \text{for a closed modal-like } \alpha.$$

Proof. The theorem is very plausible since our definitons of validity in predicate and modal cases are strictly parallel. The induction below (on the construction of formula α) only verifies this fact.

Induction base.

(a) $\alpha \equiv p$, $\alpha^x \equiv P(x)$. By the definition of the modal model corresponding to $M \equiv \langle D, V \rangle$ we have

$$\tilde{V}(p, w) = V(P)(w) = V(P(w))$$

by the definition of valuation for predicate formulas. But obviously $(P(x))_x[w] \equiv P(w)$, so

$$\tilde{V}(p, w) \equiv V(P(x)_x[w])$$

as required.

(b) $\alpha \equiv P(x)$, $\alpha^{mod} \equiv p$. We have

$$\begin{aligned} \tilde{V}(\alpha^{mod}, w) &= \tilde{V}(p, w) = V(P)(w) = V(P(w)) \\ &= V(P(x)_x[w]) \end{aligned}$$

as required.

Induction step.

Case $\vee$. (a) $\alpha \equiv \beta \vee \gamma$, so

$$\alpha^x \equiv \beta^x \vee \gamma^x \tag{3.6}$$

$$\begin{aligned} \tilde{V}(\alpha, w) &= \tilde{V}(\beta, w) \vee \tilde{V}(\gamma, w) \\ &\qquad \text{by the definition of the modal valuation} \\ &= V(\beta^x[w]) \vee V(\gamma^x[w]) \\ &\qquad \text{by the induction hypothesis} \\ &= V(\alpha^x, [w]) \\ &\qquad \text{by (3.9) and the definition of predicate} \\ &\qquad \text{valuation.} \end{aligned}$$

(b) $\alpha^{mod} \equiv \beta^{mod} \vee \gamma^{mod}$

$$\begin{aligned} \tilde{V}(\alpha^{mod}, w) &\equiv \tilde{V}(\beta^{mod}, w) \vee \tilde{V}(\gamma^{mod}, w) \\ &= V(\beta_b[w]) \vee V(\gamma_b[w]) \\ &= V((\beta \vee \gamma)_b[w]). \end{aligned}$$

Case $\neg$. Similar.

Case $\square$. $\alpha \equiv \square\beta$, and we have to prove only (3.7).

$$\begin{aligned} \tilde{V}(\alpha, w) &\equiv \min_{w_1 \in D} \tilde{V}(\beta, w_1) \\ &= \min_{w_1 \in D} V(\beta_x^x[w_1]) \\ &= V(\forall x \beta^x) = V(\alpha^x). \end{aligned}$$

Case $\forall$. $\alpha \equiv \forall x \beta$ and so is closed, since α is modal-like.

$$
\begin{aligned}
\tilde{V}(\alpha^{mod}, w) &\equiv \tilde{V}(\Box \beta^{mod}, w) \\
&= \min_{w_1 \in D} \tilde{V}(\beta^{mod}, w_1) \\
&= \min_{w_1 \in D} V(\beta_x[w_1]) \\
&= V(\forall x \beta)
\end{aligned}
$$

Case $\exists$. Similar to $\forall$.

This concludes the proof of Theorem 3.2. $\quad\square$

We are now in a position to show that we have a complete refutation procedure and a complete deductive system for S5 by simply transfering results from monadic predicate logic.

3.4 Indexed Formulas and Deduction Rules

Let us extend the definition of a valuation to the formula lists by putting

$$
\tilde{V}(\Gamma, w) \equiv \tilde{V}(\wedge \Gamma, w).
$$

In other words

$$
\tilde{V}(\{\alpha_1, \ldots, \alpha_n\}, w) = 1 \text{ iff } \tilde{V}(\alpha_i, w) = 1 \text{ for all } i \leq n.
$$

Both the refutation procedure and the deductive system will be defined for objects which are more complicated than formulas, namely for pairs (called *indexed formulas*)

$$
(\alpha, i) \tag{3.7}
$$

where α is a formula, and i is a natural number. The pair (3.10) is read "formula α in the i-th world," or shorter "α in i." The truth-value of (3.10) in the model $\langle \{w_0, w_1, w_2, \ldots\}, V \rangle$ is defined by

$$
V((\alpha, i)) = V(\alpha, w_i). \tag{3.8}
$$

The notation for V is slightly abused here, since it is treated as a mondadic (one-argument) function at the left-hand side of the equation.

As in ordinary formulas, the value for the lists of indexed formulas is defined by conjunction (i.e., taking the minimum). An indexed formula is *valid in a model (frame)* if it is true (true for any valuation in that frame), and *valid* if it is valid in any frame.

Theorem 3.2 together with definition (3.11) immediately yields the following corollary.

Corollary 3.3 *For any modal model (W, V) with $W = \{w_0, w_1, \ldots\}$, any modal formula α, and any modal-like formula β we have*

$$V((\alpha, i)) = V(\alpha_x^x[w_i]) \tag{3.9}$$
$$V(\beta^{mod}, w_i) = V(\beta_b[w_i]) \tag{3.10}$$

for the free variable b of β.

3.5 Deduction Rules

Let us define a calculus for deriving, i.e., refuting lists of signed formulas using as hints relations (3.12), (3.13), and the calculus G of the previous chapter.

Calculus GS5

- Axioms: $\Gamma, (\alpha, i), (\neg\alpha, i)$
- Inference rules.

$$(\vee)\frac{\Gamma, (\alpha, i); \quad \Gamma, (\beta, i)}{\Gamma, (\alpha \vee \beta, i)} \qquad (\neg\vee)\frac{\Gamma, (\neg\alpha, i), (\neg\beta, i)}{\Gamma, (\neg(\alpha \vee \beta), i)}$$

$$(\neg\neg)\frac{\Gamma, (\alpha, i)}{\Gamma, (\neg\neg\alpha, i)}$$

$$(\Box)\frac{\Gamma, (\alpha, j), (\Box\alpha, i)}{\Gamma, (\Box\alpha, i)} \qquad (\Diamond)\frac{\Gamma, (\alpha, k)}{\Gamma, (\Diamond\alpha, i)}(k \text{ is new})$$

$$(\neg\Box)\frac{\Gamma(\neg\alpha, k)}{\Gamma, (\neg\Box\alpha, k)}(k \text{ is new}) \qquad (\neg\Diamond)\frac{\Gamma, (\neg\alpha, j)(\neg\alpha, i)}{\Gamma, (\neg\Diamond\alpha, i)}$$

In the rules ($\Diamond$), ($\neg\Box$) there is a proviso for indices similar to the proviso for variables in the quantifier rules ($\exists$), ($\neg\forall$): the index k is new, i.e., it does not occur in the conclusion of the rule. Propositional rules do not change indices.

Let us extend the connection between formulas into a connection between derivations in GS5 and derivations in the predicate system G.

Extend operation α^x to indexed formulas by putting $(\alpha, i)^x = \alpha^x{}_x[x_i]$. First, let d be a derivation in GS5. Denote by d^x the result of replacing all indexed formulas (α, i) in d, by the predicate formula $\alpha^x_x[x_i]$.

It is easy to check (by considering each rule) that d^x is a derivation in the system G. Propositional rules are preserved by our replacement, and the modal rules are turned into the corresponding predicate rules, since the proviso for indices exactly mimics the proviso for variables in the predicate rules.

Let us now extend operation α^{mod} to derivations in G. Let d be a derivation in the system G of a modal-like formula. Assume that all formulas in d are written in variables $x_1, x_2, x_3, \ldots$ This is easily achieved by renaming variables if necessary. Denote by d^{mod} the result of replacing each formula α by the indexed formula (α^{mod}, i), where x_i is the only variable which occurs free in α. If α is closed, then the index of α^{mod} is assigned by going up the tree. If α is at the bottom node then the index of α^{mod}, and of all its predecessors up the tree having the same form α^{mod}, is 0. If a closed formula α in the derivation d is transformed into the indexed formula (α^{mod}, i), and the result α_1 of the analysis of α is a closed formula, then it is transformed in (α_1^{mod}, i). This preserves all propositional inference rules, and transforms predicate rules into modal rules.

Let us summarize the preceding.

Lemma 3.4

(a) If d *is a GS5-derivation of a formula* (α, i)*, then* d^x *is the derivation of the formula* $\alpha^x[x_i]$ *in the predicate system* G*.*

(b) If d *is a derivation in* G *of the modal-like formula* α, *wherein only the variable* x_i *is free, then* d^{mod} *is the derivation of* (α^{mod}, i) *in the system* S5. *If* α *is closed then* d^{mod} *is the derivation of* $(\alpha^{mod}, 0)$.

Example 3.4 Denote by d the derivation of the formula

$$\neg(\forall x(P(x) \land Q(x)) \to (\forall x P(x) \land \forall x Q(x)))$$

which is contained in Example 2.3 of Chapter 1.

Let us construct d^{mod} using notation $\gamma \equiv (\Box(p \land q), 0)$

axiom	axiom
$\gamma, (p, 1)(q, 1), (\neg p, 1)$	$\gamma, (p, 1), (q, 1), (\neg q, 1)$

$$\frac{\gamma, (p, 1)(q, 1), (\neg p, 1)}{\gamma, (p \land q, 1), (\neg p, 1)}$$

$$\frac{\gamma, (p \land q, 1), (\neg p, 1)}{\gamma, (\neg p, 1)}$$

$$\frac{\gamma, (\neg p, 1)}{\gamma, (\neg \Box p, 0)}$$

$$\frac{\gamma, (p, 1), (q, 1), (\neg q, 1)}{\gamma, (p \land q, 1), (\neg q, 1)}$$

$$\frac{\gamma, (p \land q, 1), (\neg q, 1)}{\gamma, (\neg q, 1)}$$

$$\frac{\gamma, (\neg q, 1)}{\gamma, (\neg \Box q, 0)}$$

$$\frac{(\Box(p \land q), 0), \neg(\Box p \land \Box q, 0)}{d^{mod} : (\neg(\Box(p \land q) \to (\Box p \land \Box q)), 0))}$$

3.6   Soundness and Completeness

Now we are ready to connect semantical and syntactical properties of GS5. Recall that our deduction systems work "up to negation."

Theorem 3.5 *System* GS5 *is sound and complete. More precisely:*

(a) *If* (α, i) *is derivable in* GS5 *for some* i, *then* $(\neg \alpha, i)$ *and* $\neg \alpha$ *are valid.*

(b) *If a modal formula* α *is valid, then* $(\neg \alpha, i)$ *is derivable in* GS5 *for any* i.

The proof is provided by corresponding properties of the system G via Lemma 3.4 and Theorem 3.3.

Proof.

(a) If d is a derivation of the formula (α, i) in GS5 then d^x is a derivation of the formula $\alpha^x[x_i]$ in G, so $\neg \alpha^x[x_i]$ is valid, and by (3.12), $(\neg\alpha, i)$ is valid. In predicate logic $\neg\alpha^x[x_i]$ implies $\neg\alpha^x{}_x[x]$ by substitution, i.e., $\neg\alpha^x$ is valid, so $\neg\alpha$ is valid as well.

(b) If a modal formula α is valid, then by (3.12) $\alpha^x[x_0]$ is valid, so α^x and $\alpha^x[x_i]$ is valid, hence $\neg\alpha^x$ and $\neg\alpha^x[x_i]$ is derivable in G. So by the previous lemma, $\neg\alpha \equiv \neg\alpha^{xmod}$ and $(\neg\alpha, i)$ is derivable in GS5 as required.   □

The previous results suggest a refutation procedure for GS5. It consists simply in starting with the formula $(\neg\alpha, 0)$ and applying all the rules of GS5 bottom-up with the same order and restrictions as in the procedure $\mathcal{P}$ for the predicate logic. Propositional rules are applied first, then rules $\Diamond$, $\neg\Box$, then the rules $\Box$, $\neg\Diamond$. In the latter two rules the index j of the new formula $(\pm\alpha, j)$ is to be chosen among already existing ones. (The role of the initial variable for a closed formula is played by the index of the bottom formula.) The modal rules are not applied if the new formula to appear is redundant. This means that at the conclusion or below there is a formula $(\pm\alpha, i)$ with a corresponding sign, and with the same i in the case of $\Box\alpha$, $\neg\Diamond\alpha$; or the formula $(\pm\alpha, l)$ with a corresponding sign and some l, in the case of $(\Diamond\alpha, \neg\Box\alpha)$.

Let us consider some examples.

Example 3.5 $p \to \Box p$
terminal, non-axiom

$$\frac{(\neg p, 1)(p, 0)}{(\neg\Box p, 0)(p, 0)} \quad \text{Introduce a new index 1 in the premise.}$$

$$\frac{}{(\neg(p \to \Box p), 0)}$$

We have the same model as in Example 3.2: there are two worlds, p is true in one of them and false in the second one.

Theorem 3.6 *The refutation procedure for S5 always terminates and is sound and complete.*

Proof. Use intertranslation between predicate and modal logic.   □

Let us now present some comments concerning the semantics of the steps of the refutation procedure.

The initial step of writing $(\neg\alpha, 0)$ corresponds to the intention of finding a countermodel of α, i.e., a model where α is false at the world 0. So $\neg\alpha$ is placed at the world 0, and it is assumed that $\neg\alpha$ is true at this world and consequences are drawn from this assumption.

The propositional steps $\mathcal{P}^{\neg\neg}, \mathcal{P}^{\vee}$, and $\mathcal{P}^{\neg\vee}$ have the same explanation as in the case of predicate logic: they reduce the truth of a formula in a world, to the truth of its components *in the same world*. Formula $\neg\neg\alpha$ is true in w iff α is true in w. Formula $\neg(\alpha \vee \beta)$ is true in w iff both formulas $\neg\alpha$ and $\neg\beta$ are true in w. Formula $\alpha \vee \beta$ is true in w iff at least one of the formulas α, β is true in w. The latter condition generates branching. Two models are possible, one where α is true, and the second where β is true. Our refutation procedure is ready to inspect both these possibilities.

Modal steps $\mathcal{P} \pm \Box$, and $\mathcal{P} \pm \Diamond$ correspond to the semantics of $\Box$ and $\Diamond$. The truth of $\Box\alpha$ in the world w implies the truth of α in every world w'. So α is placed in every world which comes into consideration if $\Box\alpha$ happens to occur in some world. The truth of $\neg\Box\alpha$ in w implies the existence of some world w' such that $\neg\alpha$ is true at w' The world w' is otherwise arbitrary, and this is taken into account by the proviso in the rule $(\neg\Box)$ requiring new indices. This feature is exactly parallel to the proviso for the variable in the rule $(\neg\forall)$ for the predicate logic and turns out to be complete for exactly the same reason.

Before the next examples let us write down definition of V for $\wedge$- and $\rightarrow$-formulas.

$$V(\alpha \wedge \beta, w) = 1 \quad iff \quad V(\alpha, w) = 1 \quad and \quad V(\beta, w) = 1$$
$$V(\alpha \rightarrow \beta, w) = 1 \quad iff \quad (V(\alpha, w) = 1 \; implies \; V(\beta, w) = 1)$$
$$iff \quad V(\neg\alpha, w) = 1 \quad or \quad V(\beta, w) = 1.$$

The steps of the refutation procedure for $\wedge, \rightarrow$ are exactly as before. At each step below, we show only formulas to be added to each line above the lowermost one.

Example 3.6 $\Box(p \to q) \to (\Box p \to \Box q)$

$$\frac{(\neg p, 1)(p, 1) \qquad (q, 1)(\neg q, 1)}{(p \to q, 1)}$$

$$\frac{}{(p, 1)}$$

$$\frac{}{(\neg q, 1)}$$

$$\frac{(\Box(p \to q), 0), (\Box p, 0), (\neg \Box q, 0)}{(\neg(\Box(p \to q) \to (\Box p \to q)), 0)}$$

The tree is closed, so the formula is valid.

Example 3.7 $\Diamond p \to \Box \Diamond p$ which is translated as

$$\exists x P(x) \to \forall x \exists x P(x)$$

$$\frac{(p, 2) \qquad (\neg p, 2)}{(p, 2)}$$

$$\frac{}{(\neg \Diamond p, 1)}$$

$$\frac{(\Diamond p, 0), \quad (\neg \Box \Diamond p, 0)}{(\neg(\Diamond p \to \Box \Diamond p), 0)}$$

Again the tree is closed, so the formula is valid.

Example 3.8 To refute the formula $(p \leftrightarrow q) \to (\Box p \leftrightarrow \Box q)$ we construct a model by applying the refutation procedure to the list $p, q, \Box p, \neg \Box q$.

$$\frac{(p, 1), (\neg q, 1), (p, 0), (q, 0)}{(\neg q, 1)}$$

$$\frac{}{(p, 0), (q, 0), (\Box p, 0), (\neg \Box q, 0)}$$

The model consists of two worlds; p is true in both, and so $\Box p$ is valid, but q is false in world 1, so $\Box q$ is false as required.

3.7   Substitution of Equivalents

We have seen that the formula

$$(p \leftrightarrow q) \rightarrow (\Box p \leftrightarrow \Box q) \qquad (3.11)$$

is not valid in S5. The connective $\leftrightarrow$ is a kind of equality
operation in the two-valued case. Sentences A and B, satisfying
$A \leftrightarrow B$, sometimes are said to have the same *extension*, and
the relation

$$(p \leftrightarrow q) \rightarrow (\alpha[p] \leftrightarrow \alpha[q])$$

is called the *extensionality* for α relative to p. So the failure
of (3.14) shows that the modality operation of necessity is non-
extensional. Sometimes one says that this shows the *intensional*
character of this operation. In more precise semantic terms,
intensionality is manifested in the dependence of the value of a
sentence on the world in which the sentence is considered.

Nevertheless, a modified version of the equivalent replacement theorem is valid in modal logic, although it takes different form in different modal logics. Consider first the predicate
translation of the formula (3.14):

$$(p(x) \leftrightarrow q(x)) \rightarrow (\forall x p(x) \leftrightarrow \forall x q(x)).$$

It is invalid, but valid formulation is obtained by prefixing a
$\forall$-quantifier to the premise:

$$\forall x(\alpha \leftrightarrow \beta) \rightarrow (\gamma[\alpha] \leftrightarrow \gamma[\beta]) \qquad (3.12)$$

for all formulas α, β containing free at most x. This immediately
implies (for $\gamma[\alpha] \equiv \gamma_p[\alpha]$), the following proposition:

Theorem 3.7 (Equivalent Replacement)

$$\Box(\alpha \leftrightarrow \beta) \rightarrow (\gamma[\alpha] \leftrightarrow \gamma[\beta]) \qquad (3.13)$$

for all modal propositional formulas α, β, γ.

Proof. By (3.15) we have

$$\forall x(\alpha^x \leftrightarrow \beta^x) \rightarrow (\gamma^x[\alpha^x] \leftrightarrow \gamma^x[\beta^x])$$

and applying modal translation and the relation

$$\delta^{xmod} \equiv \delta,$$

we have (3.16) as required.   □

A more direct proof can be given by induction on (the construction of) formula γ. See Section 4.3 below. This theorem provides the same possibilities as were pointed out for propositional logic

$$\frac{\alpha \leftrightarrow \beta}{\gamma[\alpha] \leftrightarrow \gamma[\beta]} \qquad\qquad \frac{(\alpha \leftrightarrow \beta); \gamma[\alpha]}{\gamma[\beta]}$$

At first sight it may seem that one should prefix $\Box$ to $(\alpha \leftrightarrow \beta)$, but in fact this is redundant, since we require that $(\alpha \leftrightarrow \beta)$ should be derivable.

Lemma 3.8 *The rule*

$$\frac{\alpha}{\Box\alpha} \qquad\qquad (\Box)$$

is admissible; if α is valid, then $\Box\alpha$ is valid.

Proof. The validity of α means its truth at every world w of any modal model M. Validity of $\Box\alpha$ in M means truth of $\Box\alpha$ in every world w_1 which in its turn means truth of α in every world w, and so is equivalent to the validity of α.   □

The rule $(\Box)$ is called *necessitation* or $\Box$-introduction rule.

3.8   Reduction of Modalities

Lemma 3.9 *For any formula γ beginning with a modal sign, we have*

$$\Box\gamma \leftrightarrow \gamma \ \text{in} \ \mathsf{S5} \qquad\qquad (3.14)$$

Proof. Equivalence 3.17 immediately follows from the validity of $\Box\Box\alpha \leftrightarrow \Box\alpha$ and $\Box\Diamond\alpha \leftrightarrow \Diamond\alpha$, which is established using the refutation procedure.   □

Corollary 3.10 *Any sequence μ of $\Box, \Diamond$ is equivalent in S5 to its last member, that is, $\mu\Box p \leftrightarrow \Box p$, $\mu\Diamond p \leftrightarrow \Diamond p$. So S5 has only three modalities: $\Box p, \Diamond p$, and p, i.e., $\mu p \leftrightarrow \Box p$ or $\mu p \leftrightarrow \Diamond p$ or $\mu p \leftrightarrow p$ is derivable in S5 for any such sequence μ. Furthermore, all of $\Box p, \Diamond p, p$ are different, i.e., equivalence is not derivable.*

Use the refutation procedure to construct falsifying models. We have seen that S5 contains the following formulas:

1. Propositional tautologies
2. $\Box\alpha \to \alpha$
3. $\Box(\alpha \to \beta) \to (\Box\alpha \to \Box\beta)$
4. $\neg\Box\alpha \to \Box\neg\Box\alpha$

It is closed under the following rules

$$\frac{\alpha \quad \alpha \to \beta}{\beta} \qquad \frac{\alpha}{\Box\alpha}$$

The following proposition will be proved in Chapter 6.

Theorem 3.11 *Every theorem of* S5 *is derivable from axioms 1–4 by the rules (R).*

4

System T

4.1 Accessibility of Worlds and Valuation Rules

It is time to illustrate how modal logic treats a possibility relation other than the total one, i.e., the situation where not
every world is possible relative to any other world. It seems
natural, at least at the beginning to assume that any world is
possible relative to itself. In other words, while being in a given
situation w, it is natural to assume that this situation w is possible. If the relation "world y is possible relative to a world x"
is written $R(x, y)$ then this assumption takes the form

$$R(x, x) \tag{4.1}$$

i.e., reflexivity of the relation R.

Now the modal model is a triple $\langle W, R, V \rangle$ with W, V as
before and a binary relation R between elements of the set W.

Let us see what logic corresponds to this minimal assumption concerning the possibility relation R among the worlds. In
other words, we define truth in a modal $\langle W, V, R \rangle$ by the same
clauses as before for the modal formulas and by the following
clauses for $\Box$, $\Diamond$:

$$V(\Box\alpha, w) = 1 \quad \text{iff} \quad (\forall w_1 : Rww_1)(V(\alpha, w_1) = 1) \tag{4.2}$$
$$V(\Diamond\alpha, w) = 1 \quad \text{iff} \quad (\exists w_1 : Rww_1)(V(\alpha, w_1) = 1) \tag{4.3}$$

where $(\forall w : \gamma)\delta$ means $\forall w(\gamma \rightarrow \delta)$ and $(\exists w : \gamma)\delta$ means $\exists w$
$(\gamma \wedge \delta)$.

To put it in words: formula $\Box\alpha$ is true at the world w iff α

is true in any world w_1 that is possible relative to w, i.e., such that Rww_1 holds. Formula $\Diamond\alpha$ is true in w iff α is true in some world w_1 that is possible relative to w. Formula α is valid in $\langle W, R, V\rangle$ iff $V(\alpha, w) = 1$ for all $w \in W$.

The set of all formulas valid in terms of this semantics, in all the modal models with reflexive relation R, is called the modal logic system T. A modal frame is a pair $\langle W, R\rangle$, where R is a binary relation on the nonempty set W. The formula is valid in a frame, if it is valid in any model $\langle W, R, V\rangle$.

Let us consider some properties of the system T.

Lemma 4.1

(a) If $\alpha \equiv \beta_{r_1,\ldots,r_m}[\gamma_1,\ldots,\gamma_m]$ where β is a propositional formula containing no modal connectives, then

$$V(\alpha, w) = \beta\,[V(\gamma_1, w),\ldots,V(\gamma_n, w)]$$

for any modal model $\langle W, V, R\rangle$ and any $w \in W$.

(b) In particular, if β is a tautology, then $\beta[\gamma_1,\ldots,\gamma_n]$ is valid for any formulas $\gamma_1,\ldots,\gamma_n$.

Proof. (a) is obtained by induction on the length l of formula β. The induction base (β is a propositional variable) is trivial. Here $\alpha \equiv \gamma, \beta[V(\gamma, w)] \equiv V(\gamma, w)$ and $V(\alpha, w) \equiv \beta[V(\gamma, w)] \equiv V(\gamma, w)$. The induction step follows from the fact that V is defined for boolean connectives via the same connectives applied to the components. If, for example, $\beta \equiv \beta_1 \vee \beta_2$, we use the relation:

$$V(\delta_1 \vee \delta_2, w) = V(\delta_1 w) \vee V(\delta_2, w)$$

and the inductive assumption

$$V(\beta_i[\gamma_1,\ldots,\gamma_m], w) = \beta_i[V(\gamma_1, w),\ldots,V(\gamma_m, w)] \quad i = 1, 2$$

and obtain

$$
\begin{aligned}
V(\alpha, w) &\equiv V(\beta_1[\gamma_1,\ldots,\gamma_m], w) \vee \beta_2[\gamma_1,\ldots,\gamma_m], w) \\
&= \beta_1[V(\gamma_1, w),\ldots,V(\gamma_m, w)] \vee \beta_2[V(\gamma_1, w),\ldots,V(\gamma_m, w)] \\
&= \alpha(V(\gamma_1, w),\ldots,V(\gamma_m, w))
\end{aligned}
$$

(b) immediately follows from (a). □

Lemma 4.2 *The set of formulas valid in the given frame is closed under substitution of any formulas: if α is valid then $\alpha_r[\beta]$ is valid for any formula β.*

Proof. Let α be valid in any valuation V' on the frame $\langle W, R \rangle$, i.e.,

$$V'(\alpha, w) = 1 \tag{4.4}$$

for any $w \in W$ and any valuation V'. To prove that

$$V(\alpha_r[\beta], w') = 1 \text{ for any } w' \in W$$

and given arbitrary valuation V define the new valuation V' by

$$V'(p, w) = V(p, w) \text{ for } p \neq r; V'(r, w) = V(\beta, w) \tag{4.5}$$

for any $w \in W$ and prove by induction on the construction of γ that

$$V(\gamma_r[\beta], w) = V'(\gamma, w) \tag{4.6}$$

for any formula γ and V' defined by (4.5). Then substituting α for γ in (4.6) and using Lemma 4.1 we have the desired result. □

Note that no property of R (not even reflexivity) was used in the proof of Lemmas 4.1 and 4.2, as well as in Lemmas 4.3 and 4.4 below.

Lemma 4.3 *Formula $\Box(p \to q) \to (\Box p \to \Box q)$ is valid in any (not necessarily reflexive) frame.*

Proof. We have to prove that $V(\Box(p \to q), w) = 1$ and $V(\Box p, w) = 1$ implies $V(\Box q, w) = 1$. That is, $(\forall w_1 : Rww_1) V(p \to q, w_1) = 1$ and $(\forall w_1 : Rww_1) V(p, w_1) = 1$ implies

$$(\forall w_1 : Rww_1)V(q, w_1) = 1. \tag{4.7}$$

Take an arbitrary w_1 such that Rww_1. We have

$$V(p \to q, w_1) = 1 \text{ and } V(p, w_1) = 1$$

from the premisses in (4.7), so $V(q, w_1) = 1$ as required. □

Lemma 4.4 *The set of formulas valid in the given world is closed under the rule of modus ponens:*

$$V(\alpha \to \beta, w) = 1 \text{ and } V(\alpha, w) = 1 \text{ implies } V(\beta, w) = 1$$

This is an immediate consequence of the definition of $V(\alpha \to \beta, w)$.

Lemma 4.5 *Formula $\Box p \to p$ is valid in any modal frame.*

Proof. We have to prove $(\forall w_1 : Rww_1)V(p, w_1) = 1 \to (V(p, w) = 1)$. Assume $(\forall w_1 : Rww_1)V(p, w_1) = 1$. Instantiating $\forall w_1$ by w we have $Rww \to (V(p, w) = 1)$. Using reflexivity Rww we obtain $V(p, w) = 1$ as required.   □

Lemma 4.6 *The set of formulas valid in the given model is closed under the rule of necessitation: if α is valid, then $\Box\alpha$ is valid.*

Proof. $V(\Box\alpha, w) = 1$ iff $(\forall w_1 : Rww_1)V(\alpha, w_1) = 1$, but the validity of α is even stronger statement $\forall w_1 V(\alpha, w_1) = 1$. It turns out that these lemmas provide complete axiomatization of the theorems of the system T.   □

Theorem 4.7 *The elements of T are formulas derived directly from*

1. *tautologies*
2. $\Box(p \to q) \to (\Box p \to \Box q)$
3. $\Box p \to p$

by the rules of substitution, modus ponens and necessitation.

We shall present a proof in Chapter 6.

4.2   Indexed Formulas, the System GT, and the Refutation Procedure

We approach T from the different, semantical side. A refutation procedure will be defined dealing with *indexed formulas* (α, s). This time however, indices s will not be natural numbers, but finite sequences of natural numbers. If $s \equiv \langle i_1, \ldots, i_k \rangle$ then $s * i$ means the sequence $\langle i_1, \ldots, i_k, i \rangle$ is obtained by adding the natural number i at the end, i.e., it is the extension of the sequence s by the number i. It is convenient to think of these sequences as the nodes of the universal tree of all sequences:

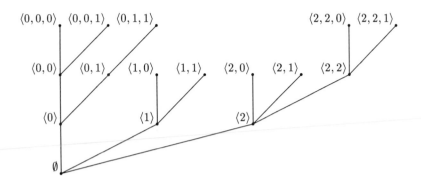

The set of all finite sequences is denoted by Seq.

At the k-th level of this tree all sequences of the length k, i.e., $\langle i_1, \ldots, i_k \rangle$ are placed in lexicograhic order. The bottom node contains the empty sequence $\emptyset$.

We define relation R on the sequences (for the system T) as follows:

$$Rss' \text{ iff } s' \equiv s \text{ or } s' \equiv s * i \text{ for some } i.$$

In other words, Rss' iff s' is either s or an immediate extension of s by one element at the end.

The refutation tree will be defined, as in the previous chapter, on the basis of a Gentzen-type system. Derivable objects of the system will be lists of the indexed formulas.

System GT

- Axioms $\Gamma, (\alpha, s), (\neg\alpha, s)$
- Inference rules.

  The *propositional rules* are the same as in GS5 with replacement (α, i) by (α, s).
- Modal rules.

$$(\Box)\frac{\Gamma, (\Box\alpha, s), (\alpha, s_1)}{\Gamma, (\Box\alpha, s)}(Rss_1) \qquad (\Diamond)\frac{\Gamma, (\alpha, s * i)}{\Gamma, (\Diamond\alpha, s)}$$

$s * i$ is new, i.e., it isn't in the conclusion.

$$(\neg\Box)\frac{\Gamma, (\neg\alpha, s * i)}{\Gamma, (\neg\Box\alpha, s)} \qquad (\neg\Diamond)\frac{\Gamma, (\neg\Diamond\alpha, s), (\neg\alpha, s_1)}{\Gamma, (\neg\Diamond\alpha, s)}(Rss_1)$$

Index s_1 in the rules $(\square), (\neg\diamond)$ is an arbitrary sequence satisfying Rss_1, i.e., at most one-element extension of s.

To save space, we often write the list $(\alpha_1, s), (\alpha_2, s), \ldots, (\alpha_m, s)$ as $(\alpha_1, \ldots, \alpha_m, s)$.

The system GT determines refutation procedure in the same way as the system GS5 in the previous chapter.

The formula α, for which we are searching for a refutation model, is placed at the bottom node $(\emptyset)$ of the tree. Each step applies some rule bottom-up to a non-closed leaf node. The leaf node is closed if it contains an explicit contradiction (α, s), $(\neg\alpha, s)$ with one and the same index (world) s. One applies first propositional rules, and only then modal rules. Among the latter, the rules $(\diamond)$, $(\neg\square)$ introducing new worlds, and putting formulas in these worlds, are to be applied first. Rules $(\square), (\neg\diamond)$ are applied lastly to all suitable worlds already present, i.e., ones satisfying $R(s, s_1)$. The restrictions ensuring termination of the procedure are that no propositional, $(\diamond)$, or $(\neg\square)$ rule is applied twice in the same branch to the same indexed formula.

Rule $(\square)$ (rule $(\neg\diamond)$) is not applied if the new indexed formula (α, s_1) (formula $(\neg\alpha, s_1)$) is already present.

The refutation procedure produces a negative result (the original formula is contradictory, i.e., does not have satisfying model) if each leaf of the refutation tree is contradictory, i.e., contains a pair $(\alpha, s), (\neg\alpha, s)$.

The refutation procedure produces a positive answer if no rule can be applied to some non-contradictory leaf L.

The modal model for this case is constructed as follows. The set W of the worlds is the set of all indices s occuring in the formulas (α, s) in L or below. The accessibility relation $R(s, s')$ means $(s' = s$ or $s' = s * i$ for some $i)$. The valuation is defined as for S5 above:

$$V(p, s) = 1 \text{ iff} (p, s) \text{ occurs in } L.$$

Example 4.1 $\alpha \equiv \Box p \rightarrow \Box\Box p$. To find a T-model for $\neg\alpha$, let us construct a refutation tree:

$$\frac{(\Box p, p, \emptyset), (p, \langle 0\rangle), (\neg p, \langle 0, 0\rangle)}{(\Box p, p, \emptyset), (\neg\Box p, p, \langle 0\rangle)}$$

$$\frac{}{(\Box p, p, \emptyset)(\neg\Box p, \langle 0\rangle)}$$

$$\frac{}{(\Box p, p, \neg\Box\Box p, \emptyset)}$$

$$\frac{(\Box p, \neg\Box\Box p, \emptyset)}{(\neg(\Box p \rightarrow \Box\Box p), \emptyset)} \qquad \text{i.e. } (\Box p, \emptyset), (\neg\Box\Box p, \emptyset)$$

The uppermost leaf cannot be extended. The only non-atomic formula there is $(\Box p, \emptyset)$, but all the possible applications of the rule $(\Box)$ are already made. One cannot add $(p, \langle 0, 0\rangle)$ since $R\emptyset, \langle 0, 0\rangle$ is false. The model is as follows, with the accessibility shown by arrows:

$$
\begin{array}{ll}
\langle 0, 0\rangle & \neg p \\
\langle 0\rangle & p, \neg\Box p \\
\emptyset & p, \Box p, \neg\Box\Box p
\end{array}
$$

We see that $\Box p$ is true at the world $\emptyset$, since p is true at $\emptyset$ and $\langle 0\rangle$. On the other hand, $\neg\Box p$ is true at $\langle 0\rangle$ since $\neg p$ is true at $\langle 0, 0\rangle$. So $\Box\Box p$ is false at $\emptyset$ as required.

Example 4.2 Construct a T-model for the formula

$$\neg(\Box(p \vee q) \rightarrow (\Box p \vee \Box q)).$$

First build up the refutation tree. Denote

$$\alpha \equiv (\Box(p \vee q), \neg\Box p, \neg\Box q, \emptyset).$$

non-closed, non-extendible, nodes

$\alpha, (\neg p, q, \langle 0 \rangle), (p, \neg q, \langle 1 \rangle), (p, \emptyset)$ $\alpha, (\neg p, q, \langle 0 \rangle), (p, \neg q, 1), (q, \emptyset)$

$\alpha, (\neg p, \langle 0 \rangle), (\neg q, 1), (p \vee q, \emptyset), (q, \langle 0 \rangle), (p, \langle 1 \rangle)$ $(\neg p, p, \langle 0 \rangle)$ *

$\alpha, (\neg p, \langle 0 \rangle), (\neg q, \langle 1 \rangle)(p \vee q, \emptyset)(p \vee q, \langle 0 \rangle), (p, \langle 1 \rangle)$ $(\neg q, q, \langle 1 \rangle)^{*}$

$\alpha, (\neg p, \langle 0 \rangle), (\neg q, \langle 1 \rangle)(p \vee q, \emptyset)(p \vee q, \langle 0 \rangle), (p \vee q, \langle 1 \rangle)$

$\alpha, (\neg p, \langle 0 \rangle), (\neg q, \langle 1 \rangle)$

$(\square(p \vee q), \neg \square p, \neg \square q, \emptyset)$

$(\square(p \vee q), \neg(\square p \vee \square q), \emptyset)$

$(\neg(\square(p \vee q) \rightarrow (\square p \vee \square q)), \emptyset)$

                            * = a contradiction

Each of the two uppermost nodes determines a model. For example, the model determined by the left hand side node is as follows.

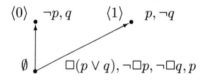

Indeed, $\square(p \vee q)$ is true at $\emptyset$ since $(p \vee q)$ is true at $\emptyset$ (for p is true at $\emptyset$), and $(p \vee q)$ is true at $\langle 0 \rangle$ (for q is true at $\langle 0 \rangle$) and at $\langle 1 \rangle$ (for p is true at $\langle 1 \rangle$). On the other hand, $\square p$ is false at $\emptyset$ since p is false at $\langle 0 \rangle$, and $\square q$ is false at at $\emptyset$ since q is false at $\langle 1 \rangle$.

As has already been pointed out, our models will usually be trees built up from indices. The depth of such an index tree is (by definition) the maximum number of nodes in one branch (not counting $\emptyset$), which for the index tree is the maximum number of elements in the index. For the tree constructed in Example 2, it is 1.

Let the modal depth $d(\beta)$ of a subformula β in α mean the number of (occurrences of) modal signs $\square, \diamond$ which have β in their scope. For example, if α is $\square \diamond p \rightarrow \square p$ then the depth of the first occurrence of p is equal to 2, and the depth of the second occurrence of p is 1.

4.3 Termination Proof, Soundness, and Completeness

Theorem 4.8 *The refutation procedure terminates for each modal formula α, with the depth of the index tree of the resulting model being at most the modal depth of subformulas in α.*

Proof. As for the system S5 above, each formula in the refutation tree for the formula α has a form $(\pm\beta, s)$ where β is a subformula of α; $\pm$ means absence or presence of the negation sign, and s is an index. We prove that length $l(s)$ of s and the depth $d(\beta)$ of β are connected. Assume that the original world is $\emptyset$. $\qquad\square$

[**Claim**] If $(\pm\beta, s)$ is present in the refutation tree, then

$$l(s) \leq d(\beta) \qquad (4.8)$$

Proof is by induction on $l(s)$. The induction base $l(s) = 0$, i.e., $s = 0$, is obvious since $0 \leq d(\beta)$.

Consider the induction step. The only steps of the reduction procedure which change $l(s)$ or $d(\beta)$ are modal steps $(\square)$, $(\neg\lozenge)$, $(\lozenge)$, $(\neg\square)$. All of them increase $d(\beta)$ by 1, and can increase $l(s)$ at most by 1, so for the new formula (β', s') one has $d(\beta') = d(\beta) + 1$ and

$$l(s') \leq l(s) + 1 \leq d(\beta) + 1 = d(\beta')$$

as required in the claim.

So for any (γ, s) in the tree we have

$$l(s) \leq d_{max} \qquad (4.9)$$

for the maximum depth d_{max} in the original formula α. So the indices s in the refutation tree are sequences of the length $\leq d_{max}$. The number of extensions $s * i$ of a given index s is bounded by the number of modal subformulas of α. Indeed, each of these extensions is made by the rule $(\lozenge)$ or $(\neg\square)$ applied to a subformula $\pm\beta$ of α, so β begins with modality, and a different i corresponds to different subformulas with $+$ or $-$ signs. So if Mod is the number of modal subformulas of α, we have

$$i \leq Mod \qquad (4.10)$$

for each i in $s * i$.

From (4.9) and (4.10) we have the estimate

$$I = Mod^{d_{max}} \qquad (4.11)$$

for the number of possible indices s.

This gives the estimate

$$SI = SMod^{d_{max}} \qquad (4.12)$$

for the number of possible signed formulas (β, s), where S is the total number of subformulas of α.

A signed formula (β, s) can be analyzed in the same branch at most $e + 1$ times where e is the number of extensions $s * i$ of the world s. This happens when β is of the form $\Box\beta'$ or $\neg\Diamond\beta'$; in all other cases (β, s) can be analyzed at most one time. But (4.10) implies $e \leq Mod$.

This gives us the estimate

$$n = SIMod$$

for the total number of steps in one branch, so the tree has at most 2^n nodes.

Theorem 4.8 is proved.

Let us now prove soundness of the refutation procedure by proving first invertibility of the rules of the system GT. To state the latter it is necessary to evaluate signed formulas in an arbitrary modal model $\langle W, R, V \rangle$. Recall that in the case of predicate logic, before evaluating a formula it was necessary to perform a substitution of individuals for the free individual variables. The modal translation α^m transforms the free individual variable x_i into an index i for S5-formulas. Then the index i is interpreted as the world w_i in enumeration of the set of worlds

$$W = \{w_0, w_1, \ldots\},$$

i.e., there is a correspondence $x_1 \mapsto w_i$.

For systems different from S5, the correspondence will be slightly more complicated.

Substitution of worlds of the frame $\langle W, R \rangle$ for the indices is any function $f \colon Seq \to W$, such that $Rss' \to Rf(s) \, f(s')$. In other words, $f(s) \in W$ for any sequence $s \in Seq$, and f preserves the accessibility of worlds.

The *value* of a signed formula (α, s) in a model $\langle W, R, V \rangle$ under the substitution f is defined by

$$V_f((\alpha, s)) \equiv V(\alpha, f(s)). \tag{4.13}$$

Theorem 4.9 *Let α be a modal formula $M = \langle W, R, V \rangle$ be a T-model such that*

$$V(\alpha, w_0) = 1 \tag{4.14}$$

for some $w_0 \in W$. Then

> *(a) there is a substitution $f : Seq \to W$ and a branch $\mathcal{B}$ of the T-refutation tree for α such that*
>
> $$V_f((\beta, s)) = 1 \tag{4.15}$$
>
> *for each signed formula $(\beta, s) \in \mathcal{B}$.*
>
> *(b) The formula $\neg\alpha$ is not derivable in GT.*

Proof. Apply bottom-up induction on the length of s to define the branch $\mathcal{B}$ (of the refutation tree or alleged proof of $\neg\alpha$ in GT) and $f(s)$ for $s \in \mathcal{B}$. The base $s = \emptyset$ is obvious: using (4.13) put

$$f(\emptyset) = w_0. \tag{4.16}$$

For the induction step consider each rule separately.

[Claim] f can be extended from the conclusion to a premiss.

For the propositional and $(\Box), (\neg\Diamond)$-steps, $f(s)$ is not changed, and for the new formulas, (4.15) follows from the induction hypothesis. For example, in the case of $(\vee)$-step applied to $(\beta_1 \vee \beta_2, s)$ we have $V_f(\beta_1 \vee \beta_2, s) = 1$ by the induction hypothesis, and so $V_f(\beta_1, s) = 1$ or $V_f(\beta_2, s) = 1$. We choose the left branch if $V_f(\beta_1, s) = 1$ and the right branch otherwise. The

$\Box$, $\neg\Diamond$ cases use the relations:

$$V(\Box\alpha, w) = 1 \wedge Rww' \rightarrow V(\alpha, w) = 1 \wedge V(\alpha, w') = 1$$

$$V(\neg\Diamond\alpha, w) = 1 \wedge Rww' \rightarrow V(\neg\alpha, w) = 1 \wedge V(\neg\alpha, w') = 1.$$

The only remaining case includes the rules $(\Diamond)$, $(\neg\Box)$. Here the new world $s * k$ is introduced to analyze formula $(\mu\beta, s)$ with $\mu \equiv \Diamond$ or $\neg\Box$. Considering for definiteness the first case and using the induction hypothesis we have $V_f(\Diamond\beta, s) = 1 = V(\Diamond\beta, f(s))$. So there is $w_1 \in W$ such that $R(f(s), w_1)$ and $V(\beta, w_1) = 1$. Choose such a w_1 and put $f(s * k) = w_1$. This will not cause conflict with the values of f for the arguments already considered since $s * k$ is the new index by the definition. Define $f(s) = w_0$ for all remaining sequences s.

Theorem 4.9(a) is proved.

(b) If $\neg\alpha$ is derivable, apply the same procedure to the derivation. $\Box$

Theorem 4.10 *The refutation procedure $\mathcal{P}$ for the system T is complete and sound.*

(a) If the refutation tree is closed then $\neg\alpha$ is valid.

(b) If t produces a model V when run for a formula α, then $V(\alpha, \emptyset) = 1$;

Proof. (a) Follows from Theorem 4.9(b).

(b) As in the case of the predicate logic we prove that if the model V was determined by the leaf l of the refutation tree, then

$$V((\beta, s)) = 1 \tag{4.17}$$

for every signed formula (β, s) occuring below node l.

Now we must prove the following lemma.

Lemma 4.11 *For any non-closed terminating branch $\mathcal{B}$ of the refutation tree the set of signed formulas (β, s) occuring in $\mathcal{B}$ is saturated, i.e.,*

(∨) *if* $((\beta \vee \gamma), s) \in \mathcal{B}$ *then* $(\beta, s) \in \mathcal{B}$ *or* $(\gamma, s) \in \mathcal{B}$.

(¬∨) *if* $(\neg(\beta \vee \gamma), s) \in \mathcal{B}$ *then* $(\neg\beta, s), (\neg\gamma, s) \in \beta$.

(¬¬) *if* $(\neg\neg\beta, s) \in \mathcal{B}$ *then* $(\beta, s) \in \mathcal{B}$.

(□) *if* $(\Box\beta, s) \in \mathcal{B}$ *then* $(\beta, s') \in \mathcal{B}$ *for every*
                                                          $s' : Rss'$ *occuring in* $\mathcal{B}$.

(¬◇) *if* $(\neg\Diamond\beta, s) \in \mathcal{B}$ *then* $(\neg\beta, s') \in \mathcal{B}$ *for every*
                                                          $s' : Rss'$ *occuring in* $\mathcal{B}$.

(◇) *if* $(\Diamond\beta, s) \in \mathcal{B}$ *then* $(\beta, s') \in \mathcal{B}$ *for some*
                                                          $s' : Rss'$ *occuring in* $\mathcal{B}$.

(¬□) *if* $(\neg\Box\beta, s) \in \mathcal{B}$ *then* $(\neg\beta, s') \in \mathcal{B}$ *for some*
                                                          $s' : Rss'$ *occuring in* $\mathcal{B}$.

Proof. By inspection of the rules.   □

Proof of Theorem 4.10 continues.

Now (4.17) is easily proved by induction on the length of the formula α. The induction base is the definition of the valuation V, and the induction step is treated by Lemma 4.11 in exactly the same way as for predicate logic.   □

Corollary 4.12 *System* GT *is sound and complete for* T*. Formula* α *is valid iff* $(\neg\alpha, \emptyset)$ *is derivable in* GT*, and iff it is valid in all finite* T*-tree models of index depth* $\leq d(\alpha)$.

Proof. Soundness. [If $(\neg\alpha, 0)$ is derivable, then α is valid] follows from Theorem 4.9(b).

Completeness. Run the T-refutation procedure for $\neg\alpha$. Since α is valid, the refutation tree will be closed by Theorem 4.10(b), and so produces the derivation of $(\neg\alpha, \emptyset)$ as required.

This concludes the proof.   □

4.4 First Degree Formulas, Substitution of Equivalents, and the Reduction of Modalities

Let us summarize our knowledge of the system T.

Theorem 4.13 *(a)* T $\subset$ S5: *if α is S5-valid, then α is T-valid.*
(b) The first degree fragments of T *and* S5 *coincide.*

Note: Here S5-validity means validity in all S5 models, i.e., when all worlds are accessible; and T-validity means belonging to the set T, i.e., validity in all models with the reflexive accessibility relation. The degree of α is the maximum number of nested modalities $\Box, \Diamond$, i.e., modalities in the scope of other modalities. A first degree formula is one without any modality in the scope of other modalities.

Proof. (a) is obvious since each S5-model is a T-model.

Another proof: each GT-derivation is transformed into GS5-derivation if we simply forget the structure of indices, i.e., enumerate the indices present in the tree

$$s_0, s_1, s_2 \ldots$$

and replace s_k by the integer k.

(b) uses the same trick, but is slightly more difficult. Let α be the first degree formula, that is to say that the modal depth of its subformulas is at most 1. By the relation between $l(s)$ and $d(\beta)$ for indexed formulas (β, s) in the refutation tree

$$l(s) \leq d(\beta) \tag{4.18}$$

we have in our case

$$l(s) \leq 1, \tag{4.19}$$

i.e., the indices can only be

$$\emptyset, \langle 0 \rangle, \langle 1 \rangle, \langle 2 \rangle, \ldots \text{ with } R(\emptyset, \langle i \rangle) \text{ for all } i*$$

Moreover, formulas β with indices $\neq \emptyset$ do not contain modalities at all. Let us identify indices $(*)$ with $0, 1, 2, \ldots$. $\Box$

[Claim] T-refutation tree for α is also an S5-refutation tree. Indeed, the S5-condition requires that the presence of $(\Box \beta, k)$ or of $(\neg \Diamond \beta, k)$ implies the presence of (β, i) or respectively of

$(\neg\beta, i)$ for all i. But in our case $k = 0$ and we have $R(0, i)$ for all i as required.

This claim implies that if $\neg\alpha$ is true in some T-model, it is true in some S5-model. So α is not S5-valid if it is not T-valid, which concludes the proof of Theorem 4.13.

Let us define strict implication: $(\alpha \Rightarrow \beta)$ is $\Box(\alpha \rightarrow \beta)$.

Corollary 4.14 *The following formulas are T-valid.*

(a) $\Box\alpha \leftrightarrow \neg\Diamond\neg\alpha;$ $\Diamond\alpha \leftrightarrow \neg\Box\neg\alpha;$
   (definitions of $\Box$, $\Diamond$ in terms of each other.)

(b) $\neg\Box\alpha \leftrightarrow \Diamond\neg\alpha;$ $\neg\Diamond\alpha \leftrightarrow \Box\neg\alpha;$
   (analogues of de Morgan's laws for $\Box, \Diamond$.)

(c) $\Box(\alpha \wedge \beta) \leftrightarrow (\Box\alpha \wedge \Box\beta);$ $\Diamond(\alpha \vee \beta) \leftrightarrow \Diamond\alpha \vee \Diamond\beta;$

   $\Box\alpha \vee \Box\beta \rightarrow \Box(\alpha \vee \beta);$ $\Diamond(\alpha \wedge \beta) \rightarrow \Diamond\alpha \wedge \Diamond\beta;$
   (distributive laws for $\Box, \Diamond$).

(d) $\alpha \Rightarrow \alpha;$ $((\alpha \Rightarrow \beta) \wedge (\beta \Rightarrow \gamma)) \Rightarrow (\alpha \Rightarrow \gamma).$
   $\Box\alpha \Rightarrow (\beta \Rightarrow \alpha)$ *but not* $\alpha \Rightarrow (\beta \Rightarrow \alpha).$
   $(\alpha \Rightarrow \beta) \wedge (\alpha \Rightarrow (\beta \rightarrow \gamma)) \Rightarrow (\alpha \Rightarrow \gamma)$ *and so* $(\alpha \Rightarrow \beta)\wedge$
   $(\alpha \Rightarrow (\beta \Rightarrow \gamma)) \Rightarrow (\alpha \Rightarrow \gamma).$
   (Properties of strict implication.)

   $(\alpha \Rightarrow (\alpha \Rightarrow \beta)) \Rightarrow (\alpha \Rightarrow \beta)$ *but not* $(\alpha \Rightarrow \beta) \Rightarrow$
   $(\alpha \Rightarrow (\alpha \Rightarrow \beta)).$

Proof. Reduce the problem to S5 by Theorem 4.13 (and necessitation) and then to the predicate logic. $\qquad\qquad\Box$

Consider further properties of strict implication. For the material implication $\rightarrow$, we have $(p \rightarrow (q \rightarrow r)) \leftrightarrow (p \wedge q) \rightarrow r$. To disprove a similar relation for $\Rightarrow$, namely $(p \wedge q \Rightarrow r) \rightarrow (p \Rightarrow (q \Rightarrow r))$, consider the predicate translation:

$$\forall x(P(x) \wedge Q(x) \rightarrow R(x)) \rightarrow \forall x(P(x) \rightarrow \forall x(Q(x) \rightarrow R(x)))$$

which is equivalent to

$$\forall x(P(x) \wedge Q(x) \rightarrow R(x)) \rightarrow (\exists x P(x) \rightarrow \forall x(Q(x) \rightarrow R(x))).$$

This formula has a countermodel

$$P(a), \neg Q(a), \neg P(b), Q(b), \neg R(b).$$

The corresponding modal model is

$$\uparrow \quad \neg p, q, \neg r$$
$$\quad \; p, \neg q$$

Let us consider the equivalent replacement theorem which in the case of S5 had the form

$$\Box(p \to q) \to (\alpha[p] \to \alpha[q]).$$

In the case of T this is no longer valid for $\alpha[p] \equiv \Box\Box p$.

$$\Box(p \leftrightarrow q) \to \Box\Box p \leftrightarrow \Box\Box q$$

is falsified at the following model (the same as for the formula $\Box q \to \Box\Box q$):

$$\uparrow \quad \neg q, p$$
$$\uparrow \quad q, \neg\Box q, p$$
$$\quad \; \Box q, p, q$$

where non-trivial R-relations are shown by vertical arrows.

The reason for the failure is the depth of the conclusion. It is easy to see that for the same reason the attempt to do the same with a fixed depth will end in failure.

$$\Box^d(p \leftrightarrow q) \to \Box^{d+1}p \leftrightarrow \Box^{d+1}q \qquad (4.20)$$

is refuted in a model of the depth $d + 1$.

The solution is to make d variable.

Theorem 4.15 *Equivalent Replacement theorem*

(a) $\Box^d(p \leftrightarrow q) \to (\alpha[p] \leftrightarrow \alpha[q])$

(b) $\Box^d(\beta \leftrightarrow \gamma) \to (\alpha[\beta] \leftrightarrow \alpha[\gamma])$ *for any formulas* $\beta, \gamma,$

where d is no less than the maximal modal depth of p in $\alpha[p]$.

Proof. Proof of (a) is by induction on the length of α. Induction base $\alpha \equiv p$ (or $\alpha \equiv r$ with $r \neq p$) is obvious. $\Box^d(p \leftrightarrow q) \to (p \leftrightarrow q)$ is the consequence of $\Box\alpha \to \alpha$, or of $\alpha \to \alpha$.

The induction step is proved in the same way as for predicate logic. Consider only the case $\alpha \equiv \Box\beta$. By the induction hypothesis we have

$$\Box^{d-1}(p \leftrightarrow q) \to (\beta[p] \leftrightarrow \beta[q]),$$

and by necessitation and $\Box(\gamma \leftrightarrow \delta) \to [\Box\gamma \leftrightarrow \Box\delta]$ this implies

$$\Box^{d}(p \leftrightarrow q) \to (\Box\beta[p] \leftrightarrow \Box\beta[q])$$

as required. (b) is obtained from (a) by substitution.   □

Consider now the number of modalities in T. Recall that the modality is any sequence of $\Box$, $\Diamond$ (including the empty one), and $\mu = v$ for two modalities in the given system if

$$\mu p \leftrightarrow vp \tag{4.21}$$

is derivable.

Theorem 4.16 *In* T *there is an infinite number of different modalities.*

$$\Box^{k}p \to \Box^{k+l}p$$

is invalid for any $k, l, > 0$.

Proof. Consider the model:

□

In fact all modalities in T are different unless they literally coincide. Prove this by means of a suitable model.

Fortunately, the formulation of the equivalent replacement theorem for provable equivalences can be simplified using necessitation rules.

Corollary 4.17 *The following rules are admissible in* T*:*

$$\frac{\beta \leftrightarrow \gamma}{\alpha[\beta] \leftrightarrow \alpha[\gamma]} \qquad\qquad \frac{\beta \leftrightarrow \gamma;\, \alpha[\beta]}{\alpha[\gamma]}$$

Proof. If $\beta \leftrightarrow \gamma$ is derivable, then by d applications of necessitation we have $\Box^d(\beta \leftrightarrow \gamma)$. Now apply Theorem 4.15 to obtain $\alpha[\beta] \leftrightarrow \alpha[\gamma]$. $\qquad\qquad\square$

Theorem 4.18 *Formula $\Box p \to p$ is valid in a frame $\langle W, R \rangle$ iff R is reflexive.*

Proof. One direction is evident. To prove another, assume R is not reflexive, i.e., not Rw_0w_0 for some w_0. Then defining $V(p, w) = 0$ iff $w = w_0$, we see $V(\Box p \to p, w_0) = 0$. $\qquad\square$

5

System S4

The last system to be considered in this course corresponds semantically to the reflexive transitive accessibility relation between worlds: (partial preorder $\leq$)

$$\forall w Rww; \quad \forall w w' w'' \, (R(w, w') \wedge R(w', w'') \to R(w, w'')). \quad (5.1)$$

These properties allow us to consider worlds as consecutive stages of discovering necessary truths.

Indeed we have

$$V(\Box\alpha, w) = 1 \to (\forall w' : Rww')V(\Box\alpha, w') = 1 \quad (5.2)$$

i.e., necessary truth is preserved under passage from a world w to an accessible world w'. Relation (5.2) is verified as follows. Let $V(\Box\alpha, w) = 1$ and $R(w, w')$. To compute $V(\Box\alpha, w')$ take arbitrary $w'' : R(w'w'')$. By (5.1) we have $R(w, w'')$, so $V(\alpha, w'') = 1$ as required.

Call a modal frame $\langle W, R \rangle$ an S4–*frame* if R is reflexive and transitive. An S4-model is a model with S4–frame. The value of a formula α is a world w, i.e., $V(\alpha, w)$ is defined for all modal models exactly as for T-models.

S4 is defined as the set of modal formulas valid in all S4-models.

Theorem 5.1

   (a) S4 *contains all instances of tautologies and is closed under substitution and necessitation.*

   (b) S4 *contains* T.

73

(c) $\Box p \to \Box\Box p$ and $\Diamond\Diamond p \to \Diamond p$ are in S4.

Proof.

(a) When the proofs were given for T, it was pointed out that they work for any modal systems defined in terms of modal models $\langle W, R, V \rangle$.

(b) is evident since every S4-model is T-model.

(c) follows from (5.2).

$\Box$

In fact, Theorem 5.1 gives axiomatization for S4, but we shall prove this later. Relations in Theorem 5.1(c) show that there are at least some reductions of modalities in S4. In fact, there is only a finite number of modalities in S4 (see Theorem 5.11).

Theorem 5.2 *Frame $\langle W, R \rangle$ is transitive iff the formula*

$$\Box p \to \Box\Box p$$

is valid on this frame.

Proof. One direction is trivial, so consider another one. Assume that R is not transitive, i.e.,

$$R(w, w'), \quad R(w', w'') \quad \text{but} \quad \neg R(w, w'').$$

Define the valuation V by putting

$$V(p, w_1) = 1 \text{ iff } R(w, w_1).$$

Then we have

$$R(\Box p, w) = \min_{Rww_1}\{V(p, w_1)\} = \min\{1\} = 1$$

and the minimum is taken over a nonempty set since $R(w, w')$ holds. Moreover, $V(p, w'') = 0$ since $\neg R(w, w'')$. So $V(\Box p, w') = 0$ since $R(w', w'')$; and $V(\Box\Box p, w) = 0$ since $R(w, w')$, as required. $\Box$

Let us describe proof rules and the refutation procedure for S4. A Gentzen-type system is denoted by GS4.

Derivable objects are again lists of signed formulas (α, s) where $s \in Seq$.

5.1 System GS4 and the Refutation Procedure

Axioms and *propositional rules* are the same as for system T, i.e., explicit contradictions and analysis of propositional connectives in the same world. *Modal inference rules* are slightly different.

For any list Γ of signed formulas, we denote by $\Gamma(s)$ the set of all formulas γ, which have index s in Γ; and by $\Gamma^\square(s)$ the set of all formulas beginning with $\square$ or $\neg\diamondsuit$ which have index s in Γ.

$$\Gamma(s) \equiv \{\gamma : (\gamma, s) \text{ is in } \Gamma\}$$
$$\Gamma^\square(s) \equiv \{\gamma : \gamma \equiv \square\gamma' \text{ or } \gamma \equiv \neg\diamondsuit\gamma' \text{ and } (\gamma, s) \text{ is in } \Gamma\}.$$

The notation (Δ, s) for $\Delta = \delta_1, \ldots, \delta_n$ means $(\delta_1, s), \ldots, (\delta_n, s)$.

Now the rules $(\square)$, $(\neg\diamondsuit)$ for GS4 are simpler than in GT, and the rules $(\diamondsuit)$, $(\neg\square)$ are slightly more complicated.

$$(\square)\frac{(\alpha, s)(\square\alpha, s), \Gamma}{(\square\alpha, s), \Gamma} \qquad \frac{(\neg\alpha, s)(\neg\diamondsuit\alpha, s), \Gamma}{(\neg\diamondsuit\alpha, s), \Gamma}(\neg\diamondsuit)$$

$$(\diamondsuit)\frac{(\alpha, \Gamma^\square(s), s*i), \Gamma}{(\diamondsuit\alpha, s), \Gamma} \qquad \frac{(\neg\alpha, \Gamma^\square(s), s*i), \Gamma}{(\neg\square\alpha, s), \Gamma}(\neg\square)$$

$$s*i \text{ is new}$$

In other words, a new world $s*i$ in the rules $(\diamondsuit), (\neg\square)$ obtains not only formula $\pm\alpha$ which it was designed to actualize, but also all $\square$-formulas from the world s. This corresponds to the persistence of $\square$-formulas noted above: all $\square$-formulas true at the world s should be also true in $s*i$. A remarkable fact is that this turns out to be sufficient to assure correct truth conditions for $\square$-formulas in all worlds.

Example 5.1 $\square p \to \square\square p$

$$\text{contradiction}$$
$$(\neg\square)\frac{(\square p, \neg\square p, \langle 0 \rangle), (\square p, \neg\square\square p, p, \emptyset)}{(\square p, \neg\square\square p, p, \emptyset)}$$
$$(\square)\frac{}{(\square p, \neg\square\square p, \emptyset)}$$

Example 5.2 Let us provide the derivation of the formula

$$\Box\Diamond p \leftrightarrow \Box\Diamond\Box\Diamond p.$$

by proving in GS4 $(\Box\Diamond p, \neg\Box\Diamond\Box\Diamond p, \emptyset)$ and $(\Box\Diamond\Box\Diamond p, \neg\Box\Diamond p, \emptyset)$. The general strategy of the proof search is to try to include more $\Box$-formulas in the current world, since they are preserved in subsequent worlds.

Example 5.3

$$
\begin{array}{c}
\text{contradiction} \\
\dfrac{(\Box\Diamond p, \neg\Box\Diamond p, \langle 0\rangle)}{(\Box\Diamond p, \neg\Diamond\Box\Diamond p, \langle 0\rangle)} \; (\neg\Diamond) \\[2ex]
\dfrac{}{((\Box\Diamond p, \neg\Box\Diamond\Box\Diamond p)\emptyset)} \; (\neg\Box)
\end{array}
$$

$$
\begin{array}{c}
\text{contradiction} \\
\dfrac{(\Diamond p, \neg\Diamond p, \langle 0,0\rangle)}{(\Box\Diamond p, \neg\Diamond p, \langle 0,0\rangle)} \\[2ex]
\dfrac{}{(\Diamond\Box\Diamond p, \neg\Diamond p, \langle 0\rangle)} \\[2ex]
\dfrac{(\Box\Diamond\Box\Diamond p, \neg\Diamond p, \langle 0\rangle)}{(\Box\Diamond\Box\Diamond p, \neg\Box\Diamond p, \emptyset)} \; (\neg\Box)
\end{array}
$$

We write $s \subset s'$ for $s, s' \in Seq$ if s' is a proper extension of s, i.e., $s' \equiv s * i_1 * \ldots * i_k, k \geq 1$.

The refutation procedure for S4 is defined now similarly to one for S5 and T. The formula to be satisfied is analyzed bottom-up according to the rules of the system.

The propositional rules are applied first, then $(\Box)$, $(\neg\Diamond)$. All these rules do not change the index of the formula.

Then the $(\Diamond)$ and $(\neg\Box)$ rules are applied to produce new worlds and the procedure is iterated.

A branch is closed if a contradiction is obtained.

Restriction: No signed formula is analyzed twice, and the rule $(\Diamond)$, is not applied to a formula $(\Diamond\alpha, s)$ if there is an s' such that $s \subseteq s'$ and all formulas $(\alpha, \Gamma^\Box(s), s')$ are present on the given branch. Similar restrictions hold for $(\neg\Box\alpha, s)$ and $(\neg\alpha, \Gamma^\Box(s), s')$.

The model extraction procedure for the refutation procedure is more complicated and will be described later.

Theorem 5.3 *The refutation procedure terminates.*

Proof. Let Mod be the number of modal subformulas of the formula to be tested. Then the set $(\alpha, \Gamma^\Box(s))$ contains at most $Mod + 1$ formula(s), and so there are at most

$$2^{Mod+1}$$

of such sets. So no formula $(\Diamond\alpha, s)$, $(\neg\Box\alpha, s)$ is analyzed to produce a new world unless

$$lth(s) \leq 2^{Mod+1} \equiv L.$$

So the total number of worlds has a bound

$$\#W \leq (Mod)^L + 1$$

since each world can have at most Mod of immediate successors (as in the case of S5 and T). So the total number of indexed formulas has a bound

$$\#(\alpha, s) \leq 2S\#W.$$

Since each signed formula is analyzed only one time in a branch, this gives the bound for the height of a branch.

The theorem is proved. $\Box$

Let us describe the *model-extraction algorithm*.

Let $\mathcal{B}$ be a finished, non-contradictory branch in a refutation tree.

Let us define the relation Rss' on the indices s, s' which occur in $\mathcal{B}$ in terms of immediate accessibility relation $R_0 s, s'$ where $R_0(s, s')$ iff $s \subseteq s'$ or s contains a formula of the form $\Diamond\alpha$ (or $\neg\Box\alpha$) which was not analyzed in the refutation tree with reference to the fact that s' contains $\alpha, \Gamma^\Box(s)$ (respectively $\neg\alpha$, $\Gamma^\Box(s)$). If the second disjunct holds, we say that s' is a $\Diamond$-successor of s.

Define R as the transitive closure of R_0, i.e., $R(s, s')$ iff there are $s_0, \ldots, s_n$ such that

$$s \equiv s_0, \quad s' \equiv s_n, \text{ and } R_0\,(s_i, s_{i+1}) \text{ for all } 0 \leq i \leq n. \qquad (5.3)$$

Put, as in the case of T and S5,

$$W \equiv \{s : s \text{ occurs in } \mathcal{B}\}$$
$$V(p, s) = 1 \text{ iff } (p, s) \text{ occurs in } \mathcal{B}.$$

The relation R is reflexive since R_0 is such, and transitive by the definition.

Lemma 5.4 *For the branch $\mathcal{B}$ as defined above:*

(a) $(\alpha \vee \beta, s) \in \mathcal{B} \rightarrow (\alpha, s) \in \mathcal{B}$ or $(\beta, s) \in \mathcal{B}$

(b) $(\neg(\alpha \vee \beta), s) \in \mathcal{B} \rightarrow (\neg\alpha, s) \in \mathcal{B}$ or $(\neg\beta, s) \in \mathcal{B}$

(c) $(\neg\neg\alpha, s) \in \mathcal{B} \rightarrow (\alpha, s) \in \mathcal{B}$

(d) $(\Diamond\alpha, s) \in \mathcal{B} \rightarrow (\exists s' : Rss')((\alpha, s) \in \mathcal{B})$
and similarly for $\neg\Box\alpha$

(e) $(\Box\alpha, s) \in \mathcal{B} \rightarrow (\forall s'(Rss' \rightarrow (\Box\alpha, s') \in \mathcal{B} \wedge (\alpha s') \in \mathcal{B})$
and similarly for $\neg\Diamond\alpha$.

Proof. (a)—(d) are obvious as before. (e) is the main clause. Let $(\Box\alpha, s) \in \mathcal{B}$ and $R(s, s')$.

Then

$$(\Box\alpha, s') \in \mathcal{B} \qquad (5.4)$$

is proved by induction on n in (5.3), i.e., on the number of R_0-steps leading from s to s'. If $n = 0$, i.e., $s \equiv s'$, relation (5.4) is obvious.

Let $(\Box\alpha, s_i) \in \mathcal{B}$ and $R_0(s_i, s_{i+1})$ (induction assumption). Consider two cases in the definition of R_0. If $s_i \subseteq s_{i+1}$, then each $\Box$-formula including $\Box\alpha$ is transferred from s_i to s_{i+1} at each $\Diamond$-step (since s_{i+1} is obtained from s_i by a series of such steps).

If s_{i+1} contains $\Gamma^\Box(s_i)$, then again it contains $\Box\alpha$. This concludes the induction step for proving (5.4) and so the proof of the first conjunct in (e).

The second conjunct follows by ($\Box$)-step.

The lemma is proved. □

Theorem 5.5 *The refutation procedure is sound and complete. The system* GS4 *is sound and complete. Every non-contradictory formula has a model on a frame* (W, R) *containing a bottom world* w, *i.e., a world satisfying* Rww' *for all* $w' \in W$.

Proved exactly as in the case of T. Existence of the bottom world is seen from the construction of the model corresponding to a refutation tree.

Example 5.4 $\Diamond \Box p \to \Box p$ is not in S4.

Example 5.5 $\Box \Box p \lor \Box \neg \Box p$ is not in S4.

Theorem 5.6 *Equivalent replacement theorem.*

$$\Box(\alpha \leftrightarrow \beta) \to (\gamma[\alpha] \leftrightarrow \gamma[\beta]) \tag{5.5}$$

Proof. Let d be the modal depth of α in $\gamma[\alpha]$. Then by the equivalent replacement theorem for T and the fact that S4 contains T we have

$$\Box^d(\alpha \leftrightarrow \beta) \to (\gamma[\alpha] \leftrightarrow \gamma[\beta]).$$

By the S4-theorems $\Box(\alpha \leftrightarrow \beta) \to \Box^d(\alpha \leftrightarrow \beta)$. Then we have (5.6) as required. ☐

5.2 The Disjunction Property, Strict Implication, and the Reduction of Modalities

Theorem 5.7 *If* $\Box \alpha_1 \lor \Box \alpha_2$ *is in S4, then* $\Box \alpha_i$ *is in S4 for some* i.

Proof. Assume by the way of contradiction, that there are S4-models $M_i = \langle W_i, R_i, V_i \rangle$ $i = 1, 2$ with bottom nodes w_1, w_2 such that

$$V_i(\alpha_i, w_i) = 0 \ , i = 1, 2.$$

Assuming that W_1, W_2 are disjoint, i.e., do not have elements in common, construct a new model by adding a new world w_0 and joining it from below to w_1 and w_2:

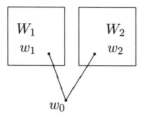

More precisely put:

$$W \equiv W_1 \cup W_2 \cup \{w_0\}$$

$$R(u,v) \equiv \begin{cases} R_i(w_i,v) & \text{if } u \equiv w_0, v \in W_i \\ true & \text{if } u \equiv w_0 \\ false & \text{otherwise} \end{cases}$$

Then R is reflexive and transitive. Put

$$V(p,w) \equiv \begin{cases} V_i(p,w) & \text{if } w \in W_i \\ 0 & \text{if } w \equiv w_0 \end{cases}$$

The value $V(\alpha, w_i)$ of any formula α depends only on values $V(\beta, u)$ of formulas β in the worlds $u \in W_i$, so

$$V(\alpha_i, w_i) = V_i(\alpha_i, w_i) = 0,$$

whence

$$V(\Diamond \neg \alpha_1 \wedge \Diamond \neg \alpha_2) w_0 = 1$$

and $\Box \alpha_1 \vee \Box \alpha_2 \leftrightarrow \neg(\Diamond \neg \alpha_1 \wedge \Diamond \neg \alpha_2)$ is false in V and cannot be in S4. $\qquad \Box$

Consider the implication fragment of S4.

Lemma 5.8 *The following formulas are S4-derivable:*

(a) $\alpha \Rightarrow \alpha$

(b) $(\alpha \Rightarrow (\beta \Rightarrow \gamma)) \Rightarrow ((\alpha \Rightarrow \beta) \Rightarrow (\alpha \Rightarrow \gamma))$

(c) $\mu \Rightarrow (\alpha \Rightarrow \mu)$ *where* μ *is modalized (i.e.,* $\equiv (\mu_1 \Rightarrow \mu_2)$*).*

Theorem 5.9 *(Without proof). (a)–(c), together with modus ponens for strict implication:*

$$\frac{A \quad A \Rightarrow B}{B}$$

constitute complete axiomatization for S4-strict implication.

Consider the modalities in S4.

We shall prove that there are exactly six non-trivial modalities in S4, and they are related in the following way:

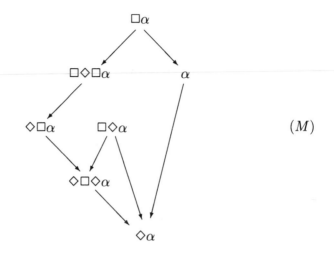

$$(M)$$

Lemma 5.10

(a) *All implications in the diagram (M) are* S4 *provable;*

(b) *none of the inverse implications is provable when* α *is a propositional letter.*

Proof. (a) Construct refutation trees, or (better) use the previously proved formulas and the replacement of equivalents. (b) Construct countermodels, for example via refutation trees. □

Theorem 5.11 *Every modality is equivalent in* S4 *to one of the modalities in* (M).

Proof. We need to consider only alternating modalities

$$\pm \Box \Diamond \Box \Diamond \ldots$$

since any two adjacent □ or ◇ can be contracted:

$$\Box\Box p \leftrightarrow \Box p; \quad \Diamond\Diamond p \leftrightarrow \Diamond p.$$

There is only one modality of depth 0, viz. p, and only two modalities of any depth greater than 0, since each such modality is completely determined by its first sign, $\Box$, or $\Diamond$. All modalities of depth $1, 2, 3$ are listed in our diagram and we have only to prove that all modalities of depth 4 is reducible to these. But we know that

$$\Box\Diamond\Box\Diamond p \leftrightarrow \Box\Diamond p$$

and so (by taking negation and substituting $\neg p$ for p)

$$\Diamond\Box\Diamond\Box p \leftrightarrow \Diamond\Box p,$$

as required. $\Box$

6

Hilbert-type Axiomatizations

We present alternative completeness proofs to show completeness of axioms systems for S5, T, S4 mentioned in Chapters 3, 4, 5 respectively. They are modelled after Makinson [1966]. The theorem of a modal system is now any formula derivable from its axioms by inference rules. Let S be a modal system (say one of S5, S4, T) containing all axioms of the classical propositional calculus (i.e., all propositional tautologies, cf. Chapter 1) and the rule of modus ponens:

$$\frac{\alpha \quad \alpha \to \beta}{\beta}.$$

We assume to simplify notation that $\Box$ is the only modal connective of S, and $\Diamond$ is defined by $\Diamond\alpha \equiv \neg\Box\neg\alpha$.

A set K of formulas will be called an L-set for S if K is consistent in S, [i.e., for any formulas $\alpha_1, \ldots, \alpha_n \in K$ the formula $\neg(\alpha_1 \wedge \ldots \wedge \alpha_n)$ is not derivable in S], and maximal, that is for any formula β one has $\beta \in K$ or $(\neg\beta) \in K$.

When S is clear from the context we talk simply of L-sets. The notion of L-set is a formal analog of the notion of possible world: all questions of truth and falsity of formulas are answered in L in a consistent way.

Lemma 6.1 *If the system* S *is consistent then for every consistent set* K *of formulas there is an* L-*set* K+ *containing* K

Proof. Let us fix enumeration $\alpha_0, \alpha_1 \ldots$ of all propositional modal formulas, and define sequence $S_0 \subseteq S_1 \subseteq \ldots$ of sets by putting: $S_0 \equiv K$;

$S_{i+1} \equiv S_i \cup \{\alpha_i\}$ *if the latter set is consistent;*

$S_{i+1} \equiv S_i \cup \{\neg\alpha_i\}$ *if $S_i \cup \{\alpha_i\}$ is inconsistent.*

Now let us define the required *L*-set by putting

$$K^+ \equiv \cup_i S_i.$$

Obviously $K^+ \supseteq S_0 \equiv K$ and for any formula α_i we have $\alpha_i \in S_{i+1} \subseteq K^+$ or $\neg\alpha_i \in S_{i+1} \subseteq K^+$ so K^+ is maximal. It remains only to prove it is consistent. If $\beta_1, \ldots, \beta_n \in S^+$ with $\neg(\beta_1 \wedge, \ldots, \wedge \beta_n)$ derivable in S, then S_i is inconsistent for i such that $\beta_1, \ldots, \beta_n \in S_i$. Let us prove by induction on i that S_i is consistent. Induction base ($i = 0$), i.e. consistency of $S_0 \equiv K$ is assumed in the theorem. Assume for induction step that S_i is consistent. If $S_i \cup \{\alpha_1\}$ is consistent, then S_{i+1} coincides with it. Let $S_i \cup \{\alpha_1\}$ be inconsistent, i.e., $\neg(\chi \wedge \alpha_1)$ is derivable in the system S for some conjunction χ of formulas in S_i. Then $S_{i+1} \equiv S_i \cup \{\neg\alpha_1\}$. Assume for contradiction that S_{i+1} is inconsistent, i.e., $\neg(\chi' \wedge \neg\alpha_1)$ is derivable in S for some conjunction χ' of formulas in S_i. Since

$$\neg(\chi \wedge \alpha_1) \wedge \neg(\chi' \wedge \neg\alpha_1) \to \neg(\chi \wedge \chi')$$

in the classical propositional logic, this implies that $\neg(\chi \wedge \chi')$ is derivable in S, but $\chi \wedge \chi'$ is a conjunction of formulas from S_i which was assumed to be consistent. This contradiction concludes the proof.   □

Lemma 6.2 *If* K *is an L-set then*

   (a) every theorem of S *is in* K;

   (b) K *is closed under modus ponens: if* $\alpha, \alpha \to \beta \in$ K *then* $\beta \in$ K;

   (c) $(\alpha \wedge \beta) \in$ K *iff* $\alpha \in$ K *and* $\beta \in$ K;

   (d) $(\alpha \vee \beta) \in$ K *iff* $\alpha \in$ K *or* $\beta \in$ K;

   (e) $(\neg\alpha) \in$ K *iff not* $\alpha \in$ K.

Proof. (a) If α is a theorem and $\neg\alpha \in K$ then K is inconsistent. Since K is maximal, one of $\alpha, \neg\alpha$ is in K, and we have $\alpha \in K$.

(b) Again by maximality one of $\beta, \neg\beta$ is in K. If it is $\neg\beta$, we have contradiction $\alpha \wedge (\alpha \to \beta) \wedge \neg\beta$.

(c) If $\alpha \wedge \beta \in K$, then neither of $\neg\alpha, \neg\beta$ is in K, since $(\alpha \wedge \beta) \wedge \neg\alpha, (\alpha \wedge \beta) \wedge \neg\beta$ are contradictions. By maximality $\alpha \in K$ and $\beta \in K$. If, on the other hand, $\alpha \in K$ and $\beta \in K$ then $\neg(\alpha\wedge\beta)$ is not in K, since $\alpha\wedge\beta\wedge\neg(\alpha\wedge\beta)$ is a contradiction. By maximality $(\alpha \wedge \beta) \in K$.

(d) Use contradictions $(\alpha \vee \beta) \wedge \neg\alpha \wedge \neg\beta, \alpha \wedge \neg(\alpha \vee \beta), \beta \wedge \neg(\alpha \vee \beta)$ in the same way as in (c).

(e) By maximality one of the formulas $\alpha, \neg\alpha$ is in K, and by consistency one of them is not in K. $\qquad\square$

Lemma 6.3 *Let modal system* S *be closed under the rule*

$$\frac{\gamma_1 \wedge \ldots \wedge \gamma_k \to \alpha}{\gamma_1 \wedge \ldots \wedge \gamma_k \to \alpha} \qquad (6.1)$$

and formula $\neg\Box\alpha$ *belongs to an L-set* K. *Then the set* K_1 *consisting of* $\neg\alpha$ *and all formulas* γ *such that* $\Box\gamma \in K$ *is consistent in* S.

Proof. Assume for contradiction that $\neg(\gamma_1 \wedge \ldots \wedge \gamma_k \neg\alpha)$ is derivable in S for some $\gamma_1, \ldots, \gamma_k$ with $\Box\gamma_1, \ldots, \Box\gamma_k \in K$. By classical propositional calculus $\gamma_1 \wedge \ldots \wedge \gamma_k \to \alpha$ is derivable in S, whence by the rule (1) one has $\Box\gamma_1 \wedge \ldots \wedge \Box\gamma_k \to \Box\alpha$ in S. By Lemma 2(c) $(\Box\gamma_1 \wedge \ldots \wedge \Box\gamma_k) \in K$, whence by Lemma 2(a), (b) one has $\Box\alpha \in K$ which contradicts $(\neg\Box\alpha) \in K$ as required. $\square$

Now we have all ingredients for constructing what is called a canonical model of the system S. In the case of S $\equiv$ S4, T this model will be characteristic for S: each non-derivable formula is refutable in one of the worlds of the model.

Definition 6.1 For a modal system S let

$$\mathcal{M}_S \equiv \langle L_S, R, V_S \rangle$$

where L_S is the family of all *L*-sets for S,

$$R(K, M) \equiv \forall\alpha(\Box\alpha \in K \text{ implies } \alpha \in M)(K, M \in L_S)$$

and V_S is the following valuation for propositional variables:

$$V_S(p, K) = 1 \equiv (p \in K).$$

Theorem 6.4 *Let* S *be one of the systems* S4, T. *Then* $\mathcal{M}_S$ *is a characteristic model for* S. *Moreover*

$$V_S(\alpha, K) = 1 \text{ iff } \alpha \in K \text{ for } K \in L_S. \tag{6.2}$$

Proof. Since all considered systems S contain axiom $\Box\alpha \to \alpha$, then R_S is reflexive by Lemma 2(a),(b): if $\Box\alpha \in K$ then $\alpha \in K$. In the case of S4 relation R_S is also transitive. Indeed let $R(K_1, K_2)$ and $R(K_2, K_3)$. If $\Box\Box\alpha \in K_1$ then by the axiom $\Box\alpha \to \Box\alpha$ and Lemma 2(a), (b) we have $\Box\Box\alpha \in K_1$, hence $\Box\alpha \in K_2$, and $\alpha \in K_3$. This implies $R(K_1, K_3)$. So any formula derivable in S is true in $\mathcal{M}_S$.

Let us prove (6.2) by induction on the construction of the formula α. The induction base (α is a propositional variable) is the definition of V_S. If α is $\alpha_1 \wedge \alpha_2$ or $\alpha_1 \vee \alpha_2$ or $\neg\alpha_1$ then use Lemma 6.2(c), (d), (e). For example

$$V_S(\alpha_1 \wedge \alpha_2, K) = 1 \leftrightarrow (V_S(\alpha_1, K) = 1 \wedge V_S(\alpha_2, K) = 1) \leftrightarrow$$

(by the induction assumption) $\alpha_1 \in K \wedge \alpha_2 \in K \leftrightarrow$ (by Lemma 2(c)) $(\alpha_1 \wedge \alpha_2) \in K$.

Consider the case $\alpha \equiv \Box\beta$. If $\Box\beta \in K$ and $R(K, M)$ then $\beta \in M$ by the definition of R. So $\Box\beta \in K$ implies $V(\Box\beta, K) = 1$. Suppose now that $\Box\beta$ is not in K. So $(\neg\Box\beta) \in K$ by completeness of K. By Lemma 6.3 the set $\{\neg\beta\} \cup K'$ where

$$K' = \{\gamma : \Box\gamma \in K\}$$

is consistent. By Lemma 6.1 there is an L-set $K_1 \supseteq K' \cup \{\neg\beta\}$. By the definition of the set K' and relation R we have $R(K, K_1)$. By consistency of K_1 formula β is not in K_1, so by the induction hypothesis $V_S(\beta, K_1) = 0$ whence $V_S(\Box\beta, K) = 0$ as required, and (6.2) is proved. □

Let now α be underivable in S. Then the set $\{\neg\alpha\}$ is consistent and by Lemma 6.1 there is an L-set $K \in L_S$ with $\neg\alpha \in K$. Then $V_S(\neg\alpha, K) = 1$, i.e., $V_S(\alpha, K) = 0$ as required.

The case of the system S5 is slightly more complicated since

$\mathcal{M}_{S5}$ is not S5-model in the sense of the Chapter 3: not every two worlds are related by the relation R. We rectify this by extracting an R-connected component.

Theorem 6.5 *Formula α is derivable in* S5 *(from axioms by inference rules) iff it is true in all* S5*-models.*

Proof. Any derivable formula is obviously true. To prove the converse take underivable formula β, so $\{\neg\beta\}$ is consistent. Put $\alpha \equiv \neg\beta$. By Lemma 6.1 there is an L-set w_0 containing α. Put

$$W = \{K : K \text{ is } L\text{-set and } Rw_0K\} \qquad (6.3)$$

i.e.,

$$K \in W \leftrightarrow \forall\gamma(\Box\gamma \in w_0 \rightarrow \gamma \in K). \qquad (6.4)$$

As in the previous theorem, put

$$V(p, K) = 1 \equiv p \in K, \qquad (6.5)$$

consider model $\langle W, V \rangle$, and prove the analog of relation (6.2) for all formulas γ:

$$V(\gamma, K) = 1 \leftrightarrow \gamma \in K. \qquad (6.6)$$

Let us note first that the relation R (defined as before) is reflexive, symmetric and transitive if the underlying system is S5. Indeed, reflexivity and transitivity were established in the proof of Theorem 6.4 since S5 contains both S4 and T. To prove symmetry assume that $R(K_1, K_2), \Box\gamma \in K_2$ and γ is not in K_1 (i.e., not $R(K_2, K_1)$). Then $\neg\gamma$ is in K_1, and by S5-axioms $\Box\gamma \rightarrow \gamma$, $\neg\Box\gamma \rightarrow \Box\neg\Box\gamma$ (and Lemma 2(a) and (b)) we have $\Box\neg\Box\gamma \in K_1$. Together with $R(K_1, K_2)$ this implies $\neg\Box\gamma \in K_2$ which contradicts $\Box\gamma \in K_2$.

Now we verify (6.6) by induction on γ The only case which is treated differently from the Theorem 6.4 is the $\Box$-case.

If $K, M \in W$ then $R(w_0, K), R(w_0, M)$ and so by symmetry and transitivity $R(K, M)$. Then $\Box\delta \in K$ implies $\delta \in M$ by the definition of R, and $V(\Box\delta, K) = 1$ follows from the induction assumption for δ and all worlds M. If δ is not in K, then as before $\neg\Box\delta \in K$ and by Lemmas 3, and 1 there is K_1 such that $R(K, K_1)$ and $\neg\delta$ is in K_1. By symmetry and transitivity of R

we have $R(w_0, K)$, so $K_1 \in W$. By the induction assumption $V(\delta, K_1) = 0$ which implies $V(\Box\delta, K) = 0$, which establishes (6.6).

The end of the proof is as in Theorem 6.4: since $\alpha \equiv \neg\beta$ is in w_0, we have $V(\neg\beta, w_0) = 1$ whence $V(\beta, w_0) = 0$ as required. $\Box$

Exercises

Each problem can be solved independently.

1. Using refutation procedure, establish validity of the following formulas in S5 and write the derivation in the system GS5.

   1.1. $\alpha \wedge (\beta \vee \gamma)) \leftrightarrow ((\alpha \wedge \beta) \vee (\alpha \wedge \gamma))$
   1.2. $(\alpha \vee (\beta \wedge \gamma)) \leftrightarrow ((\alpha \vee \beta) \wedge (\alpha \vee \gamma))$
   1.3. $\Box(\alpha \wedge \beta) \leftrightarrow (\Box\alpha \wedge \Box\beta)$
   1.4. $\Diamond(\alpha \vee \beta) \leftrightarrow (\Diamond\alpha \vee \Diamond\beta)$
   1.5. $\neg\Box\alpha \leftrightarrow \Diamond\neg\alpha$
   1.6. $\neg\Diamond\alpha \leftrightarrow \Box\neg\alpha$

2. Using the refutation procedure, construct a falsifying S5-model for the following formulas:

   2.1. $\Box(p \rightarrow q) \rightarrow (p \rightarrow \Box q)$
   2.2. $\Box(p \rightarrow q) \vee \Box(q \rightarrow p)$
   2.3. $\Box(p_1 \rightarrow p_2) \vee \Box(p_2 \rightarrow p_3) \vee \ldots \vee \Box(p_9 \rightarrow p_{10})$

3. Using standard abbreviations for $\rightarrow$, $\leftrightarrow$, De Morgan's laws

$$\neg(\alpha \vee \beta) \leftrightarrow (\neg\alpha \wedge \neg\beta),$$
$$\neg(\alpha \wedge \beta) \leftrightarrow (\neg\alpha \vee \neg\beta),$$

   relations 1.5, 1.6, and $\neg\neg\alpha \leftrightarrow \alpha$, prove that any modal formula α can be put into a *positive form* containing only $\Box$, $\Diamond$, $\wedge$, $\vee$, propositional variables and their negations ($\neg p, \neg p_1, \ldots$).
   Put each of the formulas 2.1–2.3 into the positive form.

4. Using the refutation procedure for monadic predicate logic, prove the validity of the following formulas:

$$\forall x(\gamma \lor \alpha) \leftrightarrow \gamma \lor \forall x\alpha$$
$$\exists x(\gamma \land \alpha) \leftrightarrow \gamma \land \exists x\alpha$$

provided γ does not contain x free. Construct a falsifying valuation for $\gamma \equiv P(x), \alpha \equiv Q(x)$, and some substitution for x.

5. Formula α is modalized if it begins with $\square$ or with $\diamond$, or is constructed from such formulas by propositional connectives. Using problem 4 above and the translation to the predicate logic, prove the validity of the formulas

$$\square(\gamma \lor \alpha) \leftrightarrow \gamma \lor \square\alpha$$
$$\diamond(\gamma \land \alpha) \leftrightarrow \gamma \land \diamond\alpha,$$

provided γ is modalized. Present modal countermodels for suitable non-modalized γ, α.

6. Using relations 1.1-1.4 and 5 prove that any formula in the positive form (cf. Problem 3) can be put into *miniscope form* where $\square$ preceeds only disjunctions of literals: $\square(\pm p_1 \lor \ldots \lor \pm p_n)$, and $\diamond$ preceeds only conjunctions of literals: $\diamond(\pm q_1 \lor \ldots \lor \pm q_m)$, abbreviated $\square D, \diamond K$.

7. Using relations 1.2-1.4 prove that each formula in the miniscope form (cf. Problem 6) is equivalent to a conjunction of disjunctions of $\square D, \diamond K$ for propositional D, K, and propositional D_0:

$$\alpha \leftrightarrow \wedge_i(D_{i0} \lor \square D_{i1} \lor \ldots \lor \square D_{il_i} \lor \diamond K_{i1} \lor \ldots \lor \diamond K_{im_i})$$
$$l_i, m_i \geq 0(*).$$

8. Using the refutation procedure prove that the formula of the form

$$D_0 \lor \square D_1 \lor \ldots \lor \square D_l \lor \diamond K_1 \lor \ldots \lor \diamond K_m,$$

where D_i, K_j are propositional formulas (i.e., do not contain $\square, \diamond$), is valid iff formula

$$D_i \lor (K_1 \lor \ldots \lor K_m)$$

is a propositional tautology for some $i, (0 \le i \le l)$.

9. Combine Problems 3,6,7,8 to give the procedure for deciding the validity of modal formulas. Apply it to test the validity of the formula

$$\Box((p \wedge \Diamond q) \vee (q \wedge \Diamond p)) \rightarrow$$
$$\Diamond([q \rightarrow (\Box(p \vee q) \wedge p)] \wedge [\neg q \rightarrow (\Box q \wedge p)]).$$

10. Prove that the arbitrary frame $\langle W, R \rangle$ is symmetric:

$$\forall w_0 \forall w_1 (R w_0 w_1 \rightarrow R w_1 w_0)$$

iff the formula $\Diamond \Box p \rightarrow p$ is valid there.

11. Let Br denote the logic of reflexive symmetric frame i.e., the set of formulas valid in all symmetric reflexive frames.

   11.1. Devise Gentzen-type proof rules for Br by analogy with T.

   11.2. Devise a (not necessarily terminating) refutation procedure for Br.

   11.3. Prove $\Diamond \Box p \rightarrow p$ and $\Box(\Box p \vee q) \rightarrow p \vee \Box q$ using this procedure.

   11.4. Describe the model extraction algorithm (in the case of termination).

   11.5. Prove the underivability of $\Diamond \Box p \rightarrow \Box p$, $\Box p \rightarrow \Box \Box p$. What is the relation of Br with S4, S5?

   11.6. Prove the termination of the refutation procedure. Which pattern will you follow, T or S4 and why?

12. State the equivalence replacement theorem for Br. Prove it cannot be as simple as for S4, S5.

13. What is the number of modalities in Br? Justify your answer.

CSLI Publications

Reports

The following titles have been published in the CSLI Reports series. These reports may be obtained from CSLI Publications, Ventura Hall, Stanford University, Stanford, CA 94305-4115.

Coordination and How to Distinguish Categories Ivan Sag, Gerald Gazdar, Thomas Wasow, and Steven Weisler CSLI–84–3 (*$3.50*)

Belief and Incompleteness Kurt Konolige CSLI–84–4 (*$4.50*)

Equality, Types, Modules and Generics for Logic Programming Joseph Goguen and José Meseguer CSLI–84–5 (*$2.50*)

Lessons from Bolzano Johan van Benthem CSLI–84–6 (*$1.50*)

Self-propagating Search: A Unified Theory of Memory Pentti Kanerva CSLI–84–7 (*$9.00*)

Reflection and Semantics in LISP Brian Cantwell Smith CSLI–84–8 (*$2.50*)

The Implementation of Procedurally Reflective Languages Jim des Rivières and Brian Cantwell Smith CSLI–84–9 (*$3.00*)

Parameterized Programming Joseph Goguen CSLI–84–10 (*$3.50*)

Shifting Situations and Shaken Attitudes Jon Barwise and John Perry CSLI–84–13 (*$4.50*)

Completeness of Many-Sorted Equational Logic Joseph Goguen and José Meseguer CSLI–84–15 (*$2.50*)

Moving the Semantic Fulcrum Terry Winograd CSLI–84–17 (*$1.50*)

On the Mathematical Properties of Linguistic Theories C. Raymond Perrault CSLI–84–18 (*$3.00*)

A Simple and Efficient Implementation of Higher-order Functions in LISP Michael P. Georgeff and Stephen F.Bodnar CSLI–84–19 (*$4.50*)

On the Axiomatization of "if-then-else" Irène Guessarian and José Meseguer CSLI–85–20 (*$3.00*)

The Situation in Logic–II: Conditionals and Conditional Information Jon Barwise CSLI–84–21 (*$3.00*)

Principles of OBJ2 Kokichi Futatsugi, Joseph A. Goguen, Jean-Pierre Jouannaud, and José Meseguer CSLI–85–22 (*$2.00*)

Querying Logical Databases Moshe Vardi CSLI–85–23 (*$1.50*)

Computationally Relevant Properties of Natural Languages and Their Grammar Gerald Gazdar and Geoff Pullum CSLI–85–24 (*$3.50*)

An Internal Semantics for Modal Logic: Preliminary Report Ronald Fagin and Moshe Vardi CSLI–85–25 (*$2.00*)

The Situation in Logic–III: Situations, Sets and the Axiom of Foundation Jon Barwise CSLI–85–26 (*$2.50*)

Semantic Automata Johan van Benthem CSLI–85–27 (*$2.50*)

Restrictive and Non-Restrictive Modification Peter Sells CSLI–85–28 (*$3.00*)

Institutions: Abstract Model Theory for Computer Science J. A. Goguen and R. M. Burstall CSLI–85–30 (*$4.50*)

A Formal Theory of Knowledge and Action Robert C. Moore CSLI–85–31 (*$5.50*)

Finite State Morphology: A Review of Koskenniemi (1983) Gerald Gazdar CSLI–85–32 (*$1.50*)

The Role of Logic in Artificial Intelligence Robert C. Moore CSLI–85–33 (*$2.00*)

Applicability of Indexed Grammars to Natural Languages Gerald Gazdar CSLI–85–34 (*$2.00*)

Commonsense Summer: Final Report Jerry R. Hobbs, et al CSLI–85–35 (*$12.00*)

Lecture Notes

The titles in this series are distributed by the University of Chicago Press and may be purchased in academic or university bookstores or ordered directly from the distributor: Order Department, 11030 S. Langely Avenue, Chicago, Illinois 60628.

Information-Based Syntax and Semantics. Carl Pollard and Ivan Sag. Lecture Notes No. 13. ISBN 0-937073-24-5 (paper), 0-937073-23-7 (cloth)

Non-Well-Founded Sets. Peter Aczel. Lecture Notes No. 14. ISBN 0-937073-22-9 (paper), 0-937073-21-0 (cloth)

Partiality, Truth and Persistence. Tore Langholm. Lecture Notes No. 15. ISBN 0-937073-34-2 (paper), 0-937073-35-0 (cloth)

Attribute-Value Logic and the Theory of Grammar. Mark Johnson. Lecture Notes No. 16. ISBN 0-937073-36-9 (paper), 0-937073-37-7 (cloth)

The Situation in Logic. Jon Barwise. Lecture Notes No. 17. ISBN 0-937073-32-6 (paper), 0-937073-33-4 (cloth)

The Linguistics of Punctuation. Geoff Nunberg. Lecture Notes No. 18. ISBN 0-937073-46-6 (paper), 0-937073-47-4 (cloth)

Anaphora and Quantification in Situation Semantics. Jean Mark Gawron and Stanley Peters. Lecture Notes No. 19. ISBN 0-937073-48-4 (paper), 0-937073-49-0 (cloth)

Propositional Attitudes: The Role of Content in Logic, Language, and Mind. C. Anthony Anderson and Joseph Owens. Lecture Notes No. 20. ISBN 0-937073-50-4 (paper), 0-937073-51-2 (cloth)

Literature and Cognition. Jerry R. Hobbs. Lecture Notes No. 21. ISBN 0-937073-52-0 (paper), 0-937073-53-9 (cloth)

Situation Theory and Its Applications, Vol. 1. Robin Cooper, Kuniaki Mukai, and John Perry (Eds.). Lecture Notes No. 22. ISBN 0-937073-54-7 (paper), 0-937073-55-5 (cloth)

The Language of First-Order Logic (including the Macintosh program, Tarski's World). Jon Barwise and John Etchemendy, second edition, revised and expanded. Lecture Notes No. 23. ISBN 0-937073-74-1 (paper)

Lexical Matters. Ivan A. Sag and Anna Szabolcsi, editors. Lecture Notes No. 24. ISBN 0-937073-66-0 (paper), 0-937073-65-2 (cloth)

Tarski's World. Jon Barwise and John Etchemendy. Lecture Notes No. 25. ISBN 0-937073-67-9 (paper)

Situation Theory and Its Applications, Vol. 2. Jon Barwise, J. Mark Gawron, Gordon Plotkin, Syun Tutiya, editors. Lecture Notes No. 26. ISBN 0-937073-70-9 (paper), 0-937073-71-7 (cloth)

Literate Programming. Donald E. Knuth. Lecture Notes No. 27. ISBN 0-937073-80-6 (paper), 0-937073-81-4 (cloth)

Normalization, Cut-Elimination and the Theory of Proofs. A. M. Ungar. Lecture Notes No. 28. ISBN 0-937073-82-2 (paper), 0-937073-83-0 (cloth)

Lectures on Linear Logic. A. S. Troelstra. Lecture Notes No. 29. ISBN 0-937073-77-6 (paper), 0-937073-78-4 (cloth)

A Short Introduction to Modal Logic. Grigori Mints. Lecture Notes No. 30. ISBN 0-937073-75-X (paper), 0-937073-76-8 (cloth)

Other CSLI Titles Distributed by UCP

Agreement in Natural Language: Approaches, Theories, Descriptions. Michael Barlow and Charles A. Ferguson (Eds.). ISBN 0-937073-02-4 (cloth)

Papers from the Second International Workshop on Japanese Syntax. William J. Poser (Ed.). ISBN 0-937073-38-5 (paper), 0-937073-39-3 (cloth)

The Proceedings of the Seventh West Coast Conference on Formal Linguistics (WCCFL 7). ISBN 0-937073-40-7 (paper)

The Proceedings of the Eighth West Coast Conference on Formal Linguistics (WCCFL 8). ISBN 0-937073-45-8 (paper)

The Phonology-Syntax Connection.
Sharon Inkelas and Draga Zec (Eds.)
(co-published with The University of
Chicago Press). ISBN 0-226-38100-5
(paper), 0-226-38101-3 (cloth)

*The Proceedings of the Ninth West Coast
Conference on Formal Linguistics*
(WCCFL 9). ISBN 0-937073-64-4
(paper)

Japanese/Korean Linguistics. Hajime
Hoji (Ed.). ISBN 0-937073-57-1 (paper), 0-937073-56-3 (cloth)

*Experiencer Subjects in South Asian
Languages.* Manindra K. Verma
and K. P. Mohanan (Eds.). ISBN 0-
937073-60-1 (paper), 0-937073-61-X
(cloth)

*Grammatical Relations: A Cross-
Theoretical Perspective.* Katarzyna
Dziwirek, Patrick Farrell, Errapel
Mejías Bikandi (Eds.). ISBN 0-937073-
63-6 (paper), 0-937073-62-8 (cloth)

*The Proceedings of the Tenth West Coast
Conference on Formal Linguistics*
(WCCFL 10). ISBN 0-937073-79-2
(paper)

Books Distributed by CSLI

*The Proceedings of the Third West Coast
Conference on Formal Linguistics*
(WCCFL 3). ($10.95) ISBN 0-937073-
45-8 (paper)

*The Proceedings of the Fourth West
Coast Conference on Formal Linguistics* (WCCFL 4). ($11.95) ISBN
0-937073-45-8 (paper)

*The Proceedings of the Fifth West Coast
Conference on Formal Linguistics*
(WCCFL 5). ($10.95) ISBN 0-937073-
45-8 (paper)

*The Proceedings of the Sixth West Coast
Conference on Formal Linguistics*
(WCCFL 6). ($13.95) ISBN 0-937073-
45-8 (paper)

*Hausar Yau Da Kullum: Intermediate
and Advanced Lessons in Hausa Language and Culture.* William R. Leben,
Ahmadu Bello Zaria, Shekarau B.
Maikafi, and Lawan Danladi Yalwa.
($19.95) ISBN 0-937073-68-7 (paper)

Hausar Yau Da Kullum Workbook.
William R. Leben, Ahmadu Bello
Zaria, Shekarau B. Maikafi, and
Lawan Danladi Yalwa. ($7.50) ISBN
0-93703-69-5 (paper)

Ordering Titles Distributed by CSLI

Titles distributed by CSLI may be
ordered directly from CSLI Publications, Ventura Hall, Stanford University, Stanford, California 94305-4115 or
by phone (415)723-1712 or (415)723-
1839. Orders can also be placed by e-
mail (pubs@csli.stanford.edu) or FAX
(415)723-0758.

All orders must be prepaid by
check, VISA, or MasterCard (include
card name, number, expiration date).
For shipping and handling add $2.50
for first book and $0.75 for each addi-
tional book; $1.75 for the first report
and $0.25 for each additional report.
California residents add 7% sales tax.

For overseas shipping, add $4.50
for first book and $2.25 for each addi-
tional book; $2.25 for first report and
$0.75 for each additional report. All
payments must be made in US cur-
rency.

CSLI was founded early in 1983 by researchers from Stanford University, SRI International, and Xerox PARC to further research and development of integrated theories of language, information, and computation. CSLI headquarters and the publication offices are located at the Stanford site.

CSLI/SRI International **CSLI/Stanford**   **CSLI/Xerox PARC**
333 Ravenswood Avenue   Ventura Hall   3333 Coyote Hill Road
Menlo Park, CA 94025   Stanford, CA 94305 Palo Alto, CA 94304

99 98 97 96 95 94 93 92   5 4 3 2 1

Library of Congress Cataloging-in-Publication Data

Mints, G. E.
   A short introduction to modal logic / Grigori Mints.
     p.   cm. -- (CSLI lecture notes ; no. 30)
   ISBN 0-937073-76-8
   ISBN 0-937073-75-X (pbk.)
   1. Modality (Logic). I. Title. II. Series.
BC199.M6M55 1992
160—dc20   92-2924
                                                            CIP

CSLI Lecture Notes report new developments in the study of language, information, and computation. In addition to lecture notes, the series includes monographs, working papers, and conference proceedings. Our aim is to make new results, ideas, and approaches available as quickly as possible.